ELECTROMAGNETIC PHENOMENA IN THE EARTH'S CRUST

GEOTECHNIKA 1

Selected translations of Russian
geotechnical literature

ELECTROMAGNETIC PHENOMENA IN THE EARTH'S CRUST

K.N.ABDULLABEKOV
Institute of Seismology, Uzbek Academy of Sciences, Tashkent

Translated from Russian and edited by
R.B.ZEIDLER
H*T*S, Gdańsk, Poland

A.A.BALKEMA / ROTTERDAM / BROOKFIELD / 1991

Originally published as: Elektromagnitnyye yavleniya v zemnoy korye

©1989 FAN Publishers, Uzbekistan

English translation ©1991 by A.A.Balkema, Rotterdam

ISBN 90 6191 171 0

Distributed in USA & Canada by: A.A.Balkema Publishers, Old Post Road, Brookfield, VT 05036, USA

Printed in the Netherlands

Contents

Preface

Earthquakes, volcanic eruptions, fissuring, tectonic movements and other phenomena caused by active processes in the Earth's crust, and in the lithosphere as a whole, bring about not only a formidable loss of human life, but also tremendeous material damages.

The Soviet authorities pay a lot of attention to the research, prediction and prevention of the natural calamities associated with catastrophic processes in the Earth's crust. The importance of this research has been emphasized in the programme formulated by the 27th Convention of the CPSU.

The earthquakes on the territory of Uzbekistan are the most frequent and dangerous. It was only in the years 1976–1986 that more than ten violent and destructive earthquakes with $M = 5.5$–7.3 took place. The research in the area of the contemporary dynamics of the Earth's crust, identification of the phenomena linked to the generation of earthquakes and the physical processes at the earthquake focus, along with elaboration of methods aimed at using the above phenomena in the forecast of the place, time and intensity of earthquakes are a set of the most important tasks facing geophysicists today.

Comprehensive, systematic and aim-oriented, geological and geophysical investigations of seismic hazards, and the elaboration of scientific background and forecast methods were initiated in Uzbekistan, at the Institute of Seismology of the Uzbek Academy of Sciences, soon after the destructive earthquake at Tashkent in 1966. Three geodynamical test areas have been established at Tashkent, Fergana and Kyzyl-Kum together with a network of forecast stations where routine observations were organized for a set of seismological, geophysical, hydrogeochemical, geodetical, deformational, deep-reaching and other precursory phenomena of earthquakes.

The fundamental capacity of ferromagnetics to change the magnetic properties due to elastic stresses makes it possible to employ the magnetometric method as one of primary tools in the measurement of the dynamics of elastic stresses in the Earth's crust. Extensive investigations were not conducted in that area because of the lack of apparatus of adequate accuracy. The technological background for such works in Uzbekistan was provided in the early seventies; since then the use of absolute proton magnetometers having the sensitivity up to 0.1 nT has become widespread.

Together with the magnetometric methods, the Institute for Seismology of the Uzbek Academy of Sciencies, in co-operation with the Tomsk Technical University, has put forth a progressive method for exploring the processes at the earthquake focus, employing the measurement of pulsed electromagnetic radiation in the band of 10–15 kHz.

The objective of that research was to derive relationships for the variation of electromagnetic fields associated with seismotectonic processes in the Earth's crust, and their use for exploration of the focus and elaboration of methods for earthquake prediction.

The following tasks have been postulated for the above research:

a) Elucidation of possible use of magnetometry in investigations of stresses in a seismically active region by model experiments on artificial technological objects (such as underground gas storage and water reservoirs in high mountains), and analysis of the results of laboratory studies and theoretical computations.

b) Establishment of geophysical test fields in seismically active regions of Uzbek-istan and execution of magnetometric observations by the method of repeated route-type, aeral and stationary measurements, accompanied by other methods, and encompassing the variation of pulsed electromagnetic radiation of the Earth's crust.

c) Magnetometric field observations in epicentral zones of past violent earth-quakes.

d) Identification of spatial and temporal relationships for slow and fast variation of magnetic and pulsed electromagnetic fields of the Earth, their nature and relation to regional and local seismotectonic processes in the Earth's crust.

e) Optimization of a network of geophysical observations and elaboration of methods using features of the variability for earthquake prediction.

Numerous investigations have been completed under programmes sponsored by different scientific organizations such as the Institute for the Earth's Magnetism, Ionosphere and Propagation of Radiowaves of the Soviet Academy of Sciences (IZMIRAN SSSR), the Institute for Geophysics of the Uzbek Scientific Centre of the Soviet Academy of Sciences (UNTs AN SSSR), the Institute for Earth Physics of the Soviet Academy of Sciences and the Institute of High Temperatures of the Soviet Academy of Sciences (IVTAN SSSR). A large part of routine stationary observations of the variability of electric and pulsed electromagnetic fields has been conducted since 1981 under an experimental and methodological programme co-ordinated by the Institute for Seismology of the Uzbek Academy of Sciences. All these efforts have made it possible to conduct many experimental studies in the prototype and have detected a considerable number of electromagnetic precursory effects of earthquakes.

CHAPTER 1

Stresses in the Earth's crust and their appearance in seismotectonic processes and in magnetic field changes of the Earth

1.1 PHYSICAL BACKGROUND FOR THE USE OF MAGNETOMETRY IN RESEARCH ON ELASTIC STRESSES IN THE EARTH'S CRUST

It is well known that earthquakes, much as other movements of the Earth's crust, result from the existing and time-variable stresses in the lithosphere and deeper strata of the Earth. The variation of the stresses is caused by intricate physico-chemical, thermo-dynamical and other processes occurring in the entire Earth. Many phenomena observed on the surface of the Earth result from the variation of stresses in the Earth's crust. They incorporate earthquakes, contemporary, recent and geological movements of the Earth's crust, anomalies of geophysical fields (magnetic, electric, gravitational, thermal etc), gaso-chemical composition of groundwater, the yield and level of oil, gas and groundwater, the dip and deformation of the Earth's crust etc.

'Exploration of the stress distribution in the Earth's crust and mantle is of large importance in the analysis of many basic theoretical problems of the earth sciences, and in the solution of the most fundamental practical aspects of mining and geology. It is also a prerequisite in elucidation of the causes and mechanisms of tectonic processes, identification of the energy of those processes, so little known as yet, and is required in the prediction of the place and intensity of future earthquakes...' (Gzovskiy 1975, p.403).

The stress condition of a rock massif is controlled by the gravity forces and contemporary tectonic stresses. As a rule, vertical stresses are not constant over area and differ substantially from geostatic stresses γH, in which γ = unit weight of rock and H = depth.

Until now numerous stress measurements in rock massifs have been conducted in many regions of the Soviet Union and abroad (Gzovskiy 1975).

Stress measurements in Khibinsk intrusions on the Kola Peninsula were conducted at depths of 100–600 m. The horizontal stresses σ_h are much higher than the geostatic ones: twenty times at a depth of 100 m and 4 times at 600 m. This phenomenon was also observed in rock massifs of Upper Shoria (4–6 times at a depth of 400–500 m; mines of Temirtau and Tashtagol), 5–7 times in mines of

1

Dzeskazgan, and 5–6 times at the magnetic anomaly of Kursk. In the Lvovskiy-Volynskiy region of Donbass, the vertical stresses, σ_v, were 2–4 times higher than the geostatic ones. The investigations of the stress condition of rock massifs in mines of Ingichkinsk deposit of Samarkanda region (Aripova et al. 1975) have shown that the geostatic stresses and the measured values were close to each other at a depth of 105 m while the horizontal stresses could be identified with the geostatic ones at a depth of 155 m while the vertical stresses were two times higher than their geostatic counterpart. At a depth of 130 m the quantities σ_v and σ_h were 2–5 times higher than the geostatic stresses ($\gamma H = 47.1$ kG/cm^2, $\sigma_v = \sigma_h = 126$ kG/cm^2 where 1 kG/cm$^2 = 98\ 066.5$ N/m^2). Similar measurements were also conducted in piedmont regions of northern Kuraminskiy Ridge of Kochbulakskiy deposit (Inoyatov 1982). The prevailing stress at a depth of 160 m was primarily a vertical compressive one ($\sigma_{vm} = 140$ kG/cm^2; $\gamma H = 39$ kG/cm^2; the subscript 'm' standing for 'mean') while the compressive horizontal stresses were about 55 kG/cm^2.

In the search for precursors (earthquake forerunners) and in elaboration of scientific background for earthquake prediction, one conducts routine measurements of stresses in rock in numerous geodynamical test areas. The results have shown that the stresses in the Earth's crust vary in a complex manner not only in space but also in time.

Elastic stresses bring about magnetic anisotropy of rock. The magnitude of elastic stresses cannot exceed the strength limit of rock in the range from thousands of kPa to hundreds of MPa. Kapitsa (1955) was first to study the variation of rock magnetization due to pressure and to identify the piezomagnetic factor C. It is known that the residual magnetization and magnetic susceptibility decreases along the axis of rock compression and increases in the vertical direction. From many laboratory experiments it follows that the piezomagnetic factor C is in the range of $(1–5)\ 10^{-6}$ kPa^{-1}. This corresponds to the variation of magnetization or magnetic susceptibility by a few percent with respect to the original value if the force of 10^4 kPa acts on rock (Shapiro 1966, Stacey 1964, 1965). However, there have been cases when the total magnetization deviated by 20–50% from the overall original one. This occurs if the rock with a fairly high residual magnetization is subject to considerable elastic stresses. Grabovskiy (1949, 1953), Kalashnikov & Kapitsa (1952), Bezuglaya & Skovorodkin (1970, 1971), Golovkov (1967), Vadkovskiy (1968), Maksudov (1972), Skovorodkin (1985) and Japanese researchers (Nagata 1965, Kinoshita 1968) and others have shown that the variation of magnetism depends not only on the magnitude and direction of elastic stresses but also on the type and kind of ferromagnetics, along with other factors. There is no universal law analogous to the law for reversible magnetic changes. However, one can expect that even in regions with fairly frequent earthquakes there can be substantial irreversible changes in magnetism due to viscous magnetization (Nagata & Kinoshita 1965, Bezuglaya & Maksudov 1970 and others).

A lot of attention is paid in laboratory investigations to the effect of temperature on magnetoelastic properties of rock, as the temperature under real conditions reaches hundreds of degrees at depths of 10–20 km. Investigations by Golovkov (1967) on magnetites of the Kursk magnetic anomaly have shown that the piezo-magnetic effect should not decrease substantially until the Curie temperature of ferromagnetic minerals of the rock is reached. Bezuglaya & Skovorodkin (1970) found from quantitative measurements in different rocks that the piezomagnetic factor was in the range $(2–3)10^{-6}$ kPa^{-1}.

Hence the variation of rock magnetization should not be smaller than that determined by the piezomagnetic factor C, that is from 10^{-6} kPa^{-1} or a few times more, down to the depth where the temperature is about the Curie point; substantial irreversible variation of rock magnetization can be expected in a number of cases.

Similar calculations for the variation of magnetism due to elastic stresses accompanying seismic and volcanic processes were first put forth by Stacey (1964, 1965). From an analysis of bibliographic sources dealing with the piezomagnetic effect of rock Stacey concluded that the piezomagnetic factor along the compression axis was given as follows:

$$\frac{1}{J_{0\parallel}}\frac{\partial J_{\parallel}}{\partial \sigma} = \frac{1}{\kappa_{0\parallel}}\frac{\kappa_{\parallel}}{\partial \sigma} = 1 \cdot 10^{-6} \quad kPa^{-1} \tag{1.1}$$

and along the tensile axis one has:

$$\frac{1}{J_{0\perp}}\frac{\partial J_{\perp}}{\partial \sigma} = \frac{1}{\kappa_{0\perp}}\frac{\kappa_{\perp}}{\partial \sigma} = 0.5 \times 10^{-6} \quad kPa^{-1}. \tag{1.2}$$

Figure 1 provides an example of Stacey's computations for the San Francisco area, which suffered considerably from the 1960 earthquake due to the San Andreas Fault movements. The liberated energy computed by the formula

$$E = \frac{V}{\sigma}L^2 t\frac{\sigma_0^2}{n} \tag{1.3}$$

in which
$\qquad L$ = linear dimension of focus,
$\qquad t$ = thickness of stressed layer,
$\qquad n$ = modulus of rock elasticity,
$\qquad \sigma_0$ = maximum shear stress,

was 10^{16} J.

Liberation of this energy in rock with a modulus of elasticity $n = 10^{11}$ Pa brings about elastic stress of 10^4 kPa. The rock is magnetic to the west of the fault and non-magnetic on the east side. The isolines illustrate the variation of magnetism of the Earth's crust. If for $\delta I_0 = 0.01I$ (the pressure being 10^4 kPa) one changed the magnetism by 1%, then the variation of the field (T_a) was 0.04 I. For $I = 10^{-2}$ [cgs] one has $T_a = 40$ nT while for $I = 10^{-3}$ [cgs] one obtains $T_a = 1$ nT.

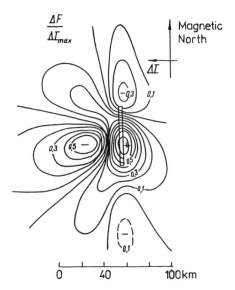

Figure 1. Computed magnetic anomaly due to accumulation of elastic stresses along active fault by Stacey (1964).

Hence the use of magnetometry in research on the dynamics of elastic stresses in the Earth's crust, together with search for earthquake precursors and investigation of processes at the earthquake focus has been based on clearcut physical (magneto-elastic effects in ferromagnetics) and geophysical (time and space variation of elastic stresses in the Earth's crust) assumptions.

1.2 LOCAL GEOMAGNETIC FIELD CHANGES

The variation of geomagnetic fields due to processes in the Earth's crust has been studied since the middle of the nineteenth century. The interest was inspired by the problem of earthquake prediction. The earliest studies described the earthquakes of 1887 (Ligurian, Equadorian and Vernenian) which were accompanied by extremely strong magnetic disturbances (Lapina 1953, Shapiro 1966, and others).

The early attempts were to link the data of the world network of magnetic observatories to the analysis of splashes on magnetographs at the time of earthquake (Lapina 1953, Shapiro 1966, and others). There was no reliable explanation for the nature of the splashes as the probability of the microphone effect was fairly high. Longer processes were not studied in terms of the data obtained from observatories since their accuracy, measured in a few nanoteslas, was insufficient.

However, in some studies (Trubjatchinskiy 1943, Omer 1946, Ol 1950, Kazanli 1948, Lakhov 1952, and others), the focuses of secular nature were compared with the active areas of the Earth's crust, although those studies were indeed below standard as the total secular course should not be compared with the processes in the Earth's crust only.

In many studies, a relationship between the field changes and the processes in the Earth's crust and the upper mantle was studied with the aid of repeated magnetic surveying. This method was used first in Japan; it consists in magnetic surveying prior to and after an earthquake or volcanic eruption. The repeated surveying enabled Krakau (1939) to identify anomalous field changes in the south of the Crimean Peninsula, and Kato & Nakamura (1934) and Kato (1935) to display a number of local segments at epicentres of violent earthquakes. However, the majority of field changes detected in the fifties were in the range of accuracy of the surveying, and therefore their reliability is doubtful (Rikitake 1968).

Since the fifties the investigation of geomagnetic field changes due to the processes in the Earth's crust has been conducted in the prototype and laboratory, and analytically. During a short time one accumulated a number of experimental and theoretical data which provided proofs for the existence of a relationship between field changes and different processes in the Earth's crust. Among the fundamental studies in this domain one should cite the laboratory investigations of the Soviet and Japanese physicists and the theoretical computations by Stacey (Kalashnikov 1954, Kapitsa 1955, Grabovskiy 1953, Kalashnikov & Kapitsa 1952, Nagata 1965, Stacey 1964, 1965).

In order to shed light on the slow variation of the geomagnetic field due to processes in the crust Orlov (1958, 1959) was first to apply repeated route observations. Those studies were initiated in 1940 on the Oka River (Orlov & Sokolov 1965) and were continued in 1947–1948 in Middle Asia (henceforth referred to as five Soviet republics of southeast Central Asia) seismoactive regions. A few routes with stations spaced every 5 km were established, and the measurements were conducted with the aid of Z-balances. As a result of the repeated observations in 1955 one identified a few segments with anomalous changes up to 40 nT. The effects were four times higher than the accuracy of measurements. The same route was passed in 1953–1962. Beginning from 1962, the H- and T-components were measured together with the Z-component of the magnetic field. The intensity of the one of the anomalies at Khabu-Rabot Pass was 100 nT over 15–17 years.

All anomalies have been associated with active deep-reaching faults. Despite the qualitative relationship between those effects and the processes in the Earth's crust it was impossible to provide any specific interpretation as the spatial and temporal picture of field variations remained unclear. It was Barsukov et al.(1968) who established later that one of the secular anomalies detected by Orlov (1958, 1959) was caused by intensive hydrochemical processes.

In recent years the method of repeated route measurements was used in many areas of the Soviet Union (Carpathians, Caucasus, Ashkhabad, Garmsk, Tashkent, Kyzyl-Kum, Fergana, Alma Ata, Baykal, etc) and abroad (the U.S.A., Japan, New Zealand etc).

In the first Middle Asian measurements one used M-2 magnetic balances and declinometers. Stable quartz magnetometers were employed later for measurements of the horizontal component. PM-1 proton magnetometers were introduced into the measurement practice in the years 1956–58, and subsequently the portable version PM-I was deployed. An important role was played by new proton magnetometers designed at the Institute for Geophysics of the Ural Division of the Soviet Academy of Sciences (Maksymovskikh & Shapiro 1976), in co-operation with the Special Design Centre of the Institute for Earth Physics of the Soviet Academy of Sciences and their Special Design Office for Physical Instrumentation of the Soviet Academy of Sciences (Tsvetkov et al. 1977 and others).

In the years 1960–1970 Pudovkin headed a group who investigated the regional and local variation of the geomagnetic field in Kamchatka. The methodology included repeated yearly measurements of the H-, Z-, D- and T-components (the T-components were measured from 1967). The spacing of the measuring points for the regional anomaly was 150–300 km, while those for local changes were spaced every 15 km. The measurements of H-, Z-, and D-components were conducted by M-2 and QHM magnetometers, while PM-5 and M-20 proton instrumentation was used for T-components. As the result one identified regional and local anomalies with the rate of 20 nT per year as for H and 40 nT per year for Z, with the accuracy of surveying of 10 nT for H and 15 nT for Pudovkin et al. (1965, 1969, 1970a, b) pointed to the correlation of the zones of secular anomalous changes and geological-tectonic regions and the intensity of secular anomaly versus the nature of volcanic activities in those regions.

Golovkov (1969, 1971) used MVS with a sensitivity of 1 nT/mm and multiple measurements of T by a proton magnetometer to discover the magnetic effect with a density of 10 nT and duration of several months. Similar measurements were carried out in the Ural Mountains (Ivanov, Shapiro et al. 1976), Kazakhstan (Finkelshtein & Kambarov 1979), Baykal (Larionov 1976, Fotiadi et al. 1970), Kirghizstan, etc.

Stationary geomagnetic observations using a single absolute magnetometer or a network thereof are the most widespread method. It has been applied in many areas of the Soviet Union, the U.S.A., New Zealand, Korea, etc.

In Middle Asia Skovorodkin (1969) discovered a magnetic effect with an intensity of several nanoteslas, which preceded the earthquake of second energy class three days before the phenomenon.

In revent years stationary observations were conducted in Carpathians, Ural Mountains, Askkhabad, Garm, Dushanbe, Tashkent, Fergana, Frunze, Alma-Ata, Baykal, and other geodynamical areas. The observations employed proton magnetometers of the following types: MPP-I, PM-OOI, APM, TMP and others. Numerous magnetic effects associated with earthquakes and other processes in the Earth' crust have been identified (Sadovskiy et al. 1979, Skovorodkin 1980, 1985, Skovorodkin & Bezuglaya 1980, Mavlanov et al. 1979a, 1983, Shapiro 1976, 1983, 1986, Yerzhanov et al. 1979, Sobakar' et al. 1975, Bezuglaya & Skovorodkin 1983, Avagimov et al. 1986, Berdaliyev 1981, Berdaliyev et al. 1986, Griaznovskaya et al.

1970, Kramarenko et al. 1969, Dembitskiy et al. 1986, Zavoyskaya & Mishchenko 1986, Larionov 1976, Muminov et al. 1986, Oganesyan et al. 1986, Maksimchuk 1983, Pyankov 1985, Kazakov 1986, Bekzhanov et al. 1972, Bushuyev 1982, and others).

In the United States, Breiner & Kovach (Breiner 1968, Breiner & Kovach 1968, Breiner 1966) investigated very fast changes in the geomagnetic field associated with movements and earthquakes in the San Andreas Fault. Five quantum magnetometers having a sensitivity of 0.02 nT/mm were deployed along the San Andreas Fault every 30 km. The variation of the field at each station was transmitted telemetrically to the central station where continuous recording of each instrument was secured individually, together with the differences between the central and individual stations. Dozens of effects associated with movements and fine earthquakes at the San Andreas Fault were identified. The intensity of those effects varried from tenths to 1 nT.

Promising results have been obtained by a group of Japanese researchers headed by Rikitake during the Matsushira earthquake swarm (Rikitake et al. 1966a, b, c, 1967, 1968). Together with the large set of geological and geophysical methods, magnetometry was also used in the course of continuous measurements of the field magnitude in the epicentral area. The seismic activity was detected twice during the investigations. The magnetic effect with an intensity of 5–7 nT corresponded to the second maximum of the seismic activity. The other occurrence of seismic activity was accompanied by intensive contemporaneous movements. The characteristic duration of the effect was 2–3 months.

In Japan there are about 100 points (stations) of the first class and 1000 points of the second class. Each point of the first class represents approximately 3000 km^2, versus 400 km^2 for the second class. Repeated surveying is conducted every 5–10 years. Tarzima (1968) provides data for the field variation throughout the entire territory of Japan over five years. He identifies a number of anomalous segments where the field variations are 3–5 nT per year. The field changes are also observed in the time domain at individual points. The intensity of some effects over 10–15 years reaches 40–50 nT. The total error of measurements was ±4 nT for the horizontal components (H), and ±0.4 nT for inclination (I) and declination (D).

Johnston & Stacey (1969) described the field changes associated with the volcanic eruption at Ngauruhoe (New Zealand) in 1968. A few hours before each eruption the magnitude of T decreased to 10 nT.

Studies on the magnetic field changes associated with seismotectonic processes in the crust have revently been intensified in China (Allen et al. 1975, Fujita 1977, Raleigh et al. 1977). Anomalous field changes with a density of 6–20 nT have been exposed by stationary measurements and repeating surveying. A report is submitted on the forecast of the Haicheng earthquake of the 4th February 1975, with M = 7.03, predicted three months before the quake, and another earthquake on 6th June 1974, with M = 4.9, predicted two days before the shock.

The primary results of geomagnetic investigations of the field dynamics observed in geodynamical test areas in the Soviet Union and abroad can be classified in the following two groups:

1. Anomalous changes in geomagnetic field associated with tectonic processes, and contemporaneous recent movements of the Earth's crust, etc. This group encompasses the anomalous changes in the Ural Mountains (Shapiro 1973), Carpathians (Kuznetsova et al. 1979, Maksimchuk 1984), Crimea (Zavoyskaya & Mishchenko 1976), Alma-Ata (Yerzhanov et al. 1979, 1982, Bushuyev 1982 and others) and in some other geodynamical test areas;

2. Anomalous field changes preceding tectonic earthquakes. Discoveries of anomalous effects in the magnetic field prior to earthquakes were very rare before 1970–1975. In the recent decade, numerous magnetic effects accompanying earthquakes have been exposed in the Soviet Union, the USA, Japan, North Korea and other countries (cf. Catalogue of Geomagnetic Precursors of Earthquakes 1984; Third All-Union Conference on Geomagnetism 1986, and others).

It can be seen from a review of field experiments that the fields of elastic stresses in the Earth's crust and the accompanying seismotectonic processes can be investigated successfully by the geomagnetic method.

1.3 SPATIAL DIMENSIONS AND CHARACTERISTIC TIMES OF TECTONIC MOVEMENTS

The spatial dimensions of tectonic structures are very different, and stretch from tenths and hundredths of millimetre to the size of large tectonic units such as platforms, geosynclinal areas, continents, oceans etc. (Belousov 1962, Khain 1973, Sadovskiy 1979, Nikolayev 1977 and others). With reference to spatial dimensions Argan & Peyve distinguished two tectonic groups, i.e. deep-reaching and crustal (Khain 1973, p.40) . The deep-reaching structures encompass practically the depths from the Earth's crust and upper mantle down to 700 km, that is the entire tectonosphere. The structures of a specific depth seem to have their typical spatial dimensions and characteristic times. With increasing depth the scale and duration increase.

Four ranks (Khain 1973, p.313) are distinguished with respect to the dimensions and duration of the growth of deep-reaching structures:

1. Deep-reaching tectonic structures, as the first-order structural elements of the lithosphere, are continents and oceans ($L = n \times 1000$ km, in which $L =$ characteristic linear dimension of the structure).

2. The second-order structural elements encompass mobile geosynclinal belts and stable blocks (platforms, epiplatform areas, plateaus). They incorporate the entire Earth's crust. This group of structures includes, for instance, the Tien Shan, Sayan, Central Atlantic Ridge, etc. ($L = n \times 1000 - n \times 100$ km).

3. Deep-reaching structural elements of the third order consist of geosynclinal systems (for instance, the Great Caucacus, southern Tien Shan, eastern Sayan etc.,

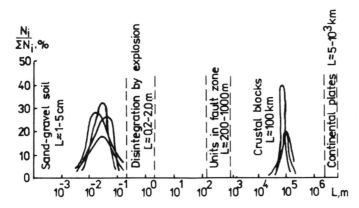

Figure 2. Distribution of natural units by their sizes, after Sadovskiy (1979).

central massifs, for instance Transcaucasian, Khankay, etc), folded systems, intra-montane, frontal and marginal depressions, shields, plateaus, etc. The third-order structures are components of the second-order structures ($L = n \times 100$ km).

4. The fourth order structures include separate depressions, folded systems (anticlinoriums, synclinoriums), platform plateaus, etc ($L = 100$–$n \times 10$ km).

The structures of the first and second order include almost the entire tectono-sphere and have global and planetary dimensions, while the third and fourth order structures encompass the entire lithosphere and have regional dimensions.

Hence the deep-reaching structures exhibit a clearcut hierarchy of spatial dimensions.

The class of crustal structures is also subdivided into ranks: 'By their scale, the crustal structures generate an entire spectrum of folded-discontinuous deformations of different orders. They can be splitted up conditionally into large (having tens or even hundreds of kilometres) and small (hundreds or tens of metres, measured along or across' (Khain 1973, p.45).

The regularity of the dimensional distribution of tectonic structures in different orders is recalled by Sadovskiy (1979). The hierarchical classification is observed from fine rock particles to large lumps, blocks and continental plateaus. The spatial dimensions of each consecutive order differ from their predecessors by a factor of 100–1000 while the internal dimensions of one order vary not more than two or three times. For instance, the dimensions of the Earth's crust blocks are in the range of 40–50 to 160 km. The sand-gravel soil has particles with characteristic size of 1–5 cm, while the dimensions of the Earth's crust blocks are about 100 km, versus those for continental plateaus about 5–10^3 km (Sadovskiy 1979) (Fig.2). The block dimensions for the Tien Shan are 80 km after Chigarev (1980), versus more than 90 km after Vilkovich et al. (1974); and those for the territories of Pamir, Minor Asia and California are about 100, 80 and 110 km, respectively.

Close figures for the dimensions of the Earth's crust blocks have also been obtained by regional investigations of anomalous geophysical fields (Tal-Virskiy

1972, 1982). The average linear dimension of the anomalies of the gravity force is 267 km, while that for the anomalies of the magnetic field reaches 310 km.

Characteristic dimensions of several orders have been identified by Rantsman (1979) for the Tien Shan, Pamir and other mountainous regions. Hierarchical structures of three orders have been postulated, for highlands (morphostructural linear element of first rank), megablocks (morphostructural linear element of second rank) and morphostructural blocks (linear element of third rank).

Hence the spatial dimensions of tectonic movements vary in very wide ranges: upon transition from low orders to high ones they obey specific rules and create a hierarchy of dimensions for tectonic movements.

Tectonic movements of the Earth are complex and embody practically all layers. The temporal spectrum of tectonic movements is particularly wide as it ranges from fractions and units of minutes to one hundred million years and more. The high-frequency band of the spectrum is characterized by many parameters such as the variation of geophysical and geochemical fields, instrumental measurements of inclinations and deformations of the Earth's crust, absolute datums of land and sea etc. Information on longer scales is provided by investigations of geological movements.

The geological processes in the Earth are cyclic. Khain (1973, p.163), with inclusion of the data of Afanasyev (1960, 1970) and Vyltsan (1967), has identified geologic cycles with characteristic times from tens and hundreds to 500–600 million years. The overall lifetime of the Earth is assessed about 5–6 million years, by different methods and data. In the Earth's history by the present time the following six cycles of tectonic activities can be distinguished: Grenvilian, Baykalian, Caledonian, Hercynian, Cimmerinian, and Alpine (Drushits & Vereshchagin 1974). The duration of the first two cycles was 300–400 million years, the middle ones lasting about 200 million years, and the recent one measuring about 150 million years.

In turn, each tectonic cycle consists of 4–5 stages of growth. For instance, each geosynclinal cycle includes five stages: 1 – early, or the stage of initial immersion; 2 – mature or preorogenic; 3 – early orogenic; 4 – orogenic proper; 5 – postorogenic or taphrogenic (Khain 1973, p.191–204).

The growth of geotectonic structures and folds occurs during a certain span of geologic time. The growth of platform folds takes tens and hundreds of million years. The duration of the development of the geosynclinal folds is counted in tens of million years. The phases of fold generation embody several million years. During some periods of the geochronological scale one encounters compaction of phases, which testifies to general increase in the intensity of folding during specific geological epochs.

'Periodicity of tectogenesis in time occurs in cycles of different duration, from many hundreds (megacycles) and first hundreds (cycles) of million years to tens of thousands, thousands, hundreds and tens of years. The short-period cycles occur as stages in long period ones' (Khain 1973, p.443).

Hence each geotectonic structure has its proper spatial dimensions and, apparently, time scales.

1.4 SPATIAL-TEMPORAL CHARACTERISTICS OF SEISMIC MOVEMENTS

The investigation of the spatial distribution of earthquakes has been described in numerous studies (Anan'in 1977, Byuss 1948, 1952, 1955, Vilkovich, Keylis-Borok et al. 1974, Gorshkov 1981, Gubin 1960, 1978, Gelfond et al. 1972, 1974, Gutenberg & Richter 1948, Ibragimov 1970,1978, Ibragimov & Abdullabekov 1974a, b, Kirillova et al. 1972a, b, Kuchay 1981, Nikolayev 1978, Nersesov et al. 1974, Petrushevskiy 1955, 1960, Rantsman 1979, Chigarev 1980, 1982, and others).

The distribution of earthquakes over the Earth globe is non-uniform. In the course of many years the focuses of earthquakes are concentrated in some specific geographic area. About 75% of earthquakes occurs in the Pacific seismic zone, 23% in the Mediterranean-Asian zone and about 2% in the remaining parts of the Earth.

An analysis of seismicity versus geological-tectonic structure shows that earthquake focuses are attributed to the following features:

1. Movable areas of the Earth's crust, that is regions of Alpine folding, old geosynclinal folds, platforms activated in recent times, segments of continental and rift zones, boundaries between lithospheric plateaus, etc (Aprodov 1965, Belousov & Gzovskiy 1954, Gorshkov & Krestnikov 1980, Stacey 1972, Khain 1973, and others).

2. Areas with high gradients of contemporary and recent movements of the Earth's crust, gravitational, magnetic, thermal and other fields (Abdullabekov et al. 1980b, Borisov 1967, Gzovskiy 1954, 1975, Ulomov 1974, and others).

3. Zones of tectonic disturbances, i.e. deep-reaching folds, overthrusts, shifts, overthrust shifts, flexural-discontinuity zones, oceanic troughs etc (Bune et al. 1960, Ulomov 1974, Zakharova 1962, Ibragimov 1970, 1978, and others).

4. Definite seismogenic zones differing by a combination of geological-tectonic, geophysical and other conditions (Gubin 1978, Ibragimov 1978, and others).

5. Morphostructural nodes (Gelfond et al. 1972, 1974, Rantsman 1979, Gorshkov et al. 1979, Vilkovich et al. 1974).

Hence the relationship of seismicity and geological-tectonic structure is obvious. From the above investigations it follows that the characteristic dimensions of areas of seismic hazard (regions of seismic zones, etc) are identical with large geostructural units of the first two ranks to which the eartquake focuses are attributed. Determination of the spatial dimensions of the subsequent ranks by the above data is difficult.

With reference to the above statements, the investigation of the spatial distribution of earthquakes harnesses at present the methods of mathematical statistics (Anan'in 1977, Gelfond et al. 1972, Vilkovich, Keylis-Borok et al. 1974, Normatov et al. 1981, 1983, Chigarev 1980, 1982, and others).

Figure 3. Chart of seismic activities of the Earth by Doubourdien (1973): (1) boundaries of continents (2) boundaries of earthquake zones with M = 8.0 (7); M = 7.9 (6); M = 7.4–7.9 (5); M = 6.9–7.4 (4); M = 6.4–6.9 (3); (3) area of seismic activity.

As the result of the investigation of earthquake distribution in the region of the Pamir Knot and the Tien Shan, Chigarev (1980,1982) found that the sequence of violent earthquakes depends on the block structure and that the subsequent earthquakes follow at distances greater than 30–40 km from the site of past earthquakes.

From the data of Anan'in (1975) it appears that earthquake focuses in specific seismogenic zones are located at distances greater than three characteristic dimensions of the focus ($R = 3r$ in which $R = $ radius of fault stress area).

Nikolayev (1978) has found that the spacing of earthquakes depends on their energy class. By analysis of the mean distances between earthquakes of different energy classes he determined the volumes responsible for the generation of earthquakes (they have been in the ranges of 40–50 and 150–200 km).

The spatial distribution of earthquakes in the years 1955–1970 in the test area of Garm were investigated by Ponomarev et al. (1976). A non-uniform aeral distribution of the epicentres was established and segments of background phenomena in their grouping have been identified. The comparison against the Poisson distribution has exposed the presence of stable deviations. The concentration of earthquake focuses coincided with segments of paleoseismic dislocations (Nersesov et al. 1974).

Non-uniform distribution of earthquake focuses is characteristic for the entire territory of the Pamir and the Tien Shan. The focuses of violent earthquakes belong to specific areas of seismic hazard, i.e. Tashkent, Leninabad, Andizhan, Namangan, Kurshab, Chatkal, Dzhanbul, etc. The places of focus concentration generate chains in sub-zonal and sub-meridian directions (Chart of Earthquake Epicentres; New Catalogue 1975). An analogous picture is observed on the Caucasian territory (Byuss 1948, 1952, 1955). The epicentral areas of Shemakhin, Alaver, Leninakan, Priyerevevansk, Gori, Apsheron and others have been singled out.

The above regularity of the distribution of earthquake focuses is typical of the entire Earth. The above statement is supported by the evidence given in the Chart of Earth Seismicity compiled by Doubourdien (1973) (Fig.3). From the chart it is clear that focuses of violent earthquakes are linked to specific areas of seismic activity having particular characteristic dimensions, and that the segments of earthquake focus concentration have also particular dimensions and spatial configuration. However, the research on the spatial distribution of earthquake focuses is restricted to qualitative description and comparison of geological-tectonic situations. A detailed analysis using contemporary computer-aided methods of mathematical statistics can be found in a few studies only.

The spatial features of the seismicity in the Caucasus, the Balkan Peninsula, Central Asia and Italy have been investigated by statistical methods dealing with distributions of events in the two-dimensional space (Nurmatov et al. 1981, 1983, Nurmatov 1984). The method of recurrential aeral averaging has been based on a filtering sequence applied to the spatial distribution of events using two-

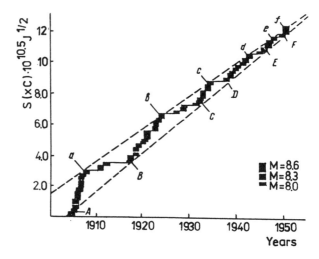

Figure 4. Benioff's (1951) graph for the Earth globe.

dimensional filters of different sizes. The spatial characteristics of the seismic activity display hierarchical structure. Five characteristic dimensions, ranging from a seismic zone to the entire surface of the Earth have been discriminated: I $= 30–40$ km; II $= 150–200$ km; III $= 800–1000$ km; IV $= n \times 1000$; V $= 4 \times 10^4$ km.

Many researchers have attempted to discriminate periodicity in the repeated earthquakes for both entire Earth and individual regions and zones (Abdullabekov 1979, Vadkovskiy et al. 1978, Gayskiy 1970, Golovkov 1977, Ibragimov & Abdullabekov 1974a, b, Kirillova 1957, Karnik 1969, Lamakin 1966, Lursmanashvili 1963, Magnitskiy 1965, Mei-Shi-Yun 1960, Mogi 1966, Rikitaki 1979, Fedotov 1968, Aki 1956, Kawasumi 1970, Utsu 1972, and others).

In the opinion of many specialists (Khain 1973, Magnitskiy 1965, Vvedenskaya 1969, Ulomov 1974, Mogi 1976, Rikitaki 1979, and others), the Earth is a unique gigantic system of stresses. A change in the stress condition at one place brings about redistribution of the elastic stresses at other places. This is supported by a diversified evidence. We shall discuss some of pertinent factors.

Basing on statistical data analysis for the most violent earthquakes with $M \geq 8$ which took place on the Earth in the years 1904-1950, Benioff (Magnitskiy 1965) has established that the seismic activity of the Earth is cyclic (Fig.4), so that there appear separate years of active outbursts of seismic energy. Magnitskiy assumes that this cyclic behaviour is not random and takes this as a proof that the most violent earthquakes are not independent. Despite the local nature of individual earthquakes they are associated with general global events altogether. Cases are known (Bott 1974) when a violent earthquake occurs at one place and the other places of the planet are calm.

In the Soviet Union, basic seismo-static studies on the spatial and temporal features of seismicity have been carried out for three regions, viz. the Caucasus,

Middle Asia and the Far East; a few works have been published for the territories of Peribaykal, Baltic shield, and some other regions.

Fedotov (1968) exposed a seismic zone displaying the periodicity of violent earthquakes with $T = 140 \pm 60$ years for the coasts of Kamchatka and the Kuril Islands of the Pacific seismic zone. A group of researchers (Fedotov et al. 1976) later conducted detailed seismostatistical investigations and exposed shorter periods having $T = 13.6$ and $T = 6.2$ years, aside from the earlier $T = 140 \pm 60$ years.

Until recently the fullest information on historical earthquakes has been collected for the Caucasus (Byuss 1948, 1952, 1955). Many studies have been carried out in the area of seismostatistics, which is worthwhile to note. There are many publications for the Middle Asia (Vilkovich, Keylis-Borok et al. 1972, Bune 1970, Ibragimov 1970, Ibragimov & Abdullabekov 1974, Ulomov 1964, Nersesov et al. 1970, Green et al. 1976, 1978, 1980, and others). The cycles with periods of 260, 110, 40, 15–20, 11, 10 and 5.56 years have been identified.

Seismostatistical investigations have also been carried out for Japan (Rikitaki 1979, Mogi 1976, Aki 1968, Kawasumi 1970, Usami & Hisomoto 1970, 1971), Chile (Magnitskiy 1965, Duda 1966, Tamrazyan 1962), People's Republic of Korea (Mei-Shi-Yun 1960), and the Baltic regions (Vadkovskiy et al. 1978, Karnik 1969 and others).

As an example we will discuss the results of investigations collected at the World Data Centre (USSR). Vadkovskiy et al. (1978) have thoroughly analysed the catalogue of violent earthquakes for the Balkan Peninsula over the time span from the 21st century B.C. to 1969. In order to identify the periodicity, regions with synchronous seismic conditions were groupped in the first stage, and then the research concentrated on the periodicity. We will consider the results for the spectrum obtained by three methods using the Bartlett window, Tuckey window and the maximum entropy (Fig.5). It is seen from the drawing that the major period of $T = 35$ years and the harmonics with $T = 17$–18 years and $T = 11$–12 years are clearly identified. The result is close to the period of 30 years given earlier by Karnik (1969) for the European region.

The statistical analysis of earthquakes has shown that the seismic energy distribution is cyclic for both the entire Earth and individual regions (Table 1). The most characteristic times appearing in many regions are the cyclic periods of 5.05–5.75 (the Caucasus and Middle Asia), 11 (the Caucasus, Middle Asia, Japan and the Balkan Peninsula), 17.5–22 (the Caucasus, the Kuril-Kamchatka zone, Baykal, Middle Asia, the Balkan Peninsula), 30–40 (the Caucasus, Middle Asia, the Balkan Peninsula, Europe and the interface of the Indo-Australian and Eurasian shields), 80, 100–140, 200–284, 575, 1100, 1500, etc years.

It is seen that the characteristic times and linear dimensions of tectonic processes and seismic activity are an outcome of the variation of the fields of elastic stresses. The spatial and temporal parameters of the electromagnetic phenomena associated with seismotectonic processes should have identical characteristic times.

Table 1. Cyclic recurrence of violent earthquakes.

Region	Identified cycles, yrs	Reference
Caucasus	1100, 500–650, 88, 22	Kirillova 1957,
		Tamrazyan 1962
	11, 5.75, 5.05, 34.5	Lursmanashvili 1973,
		Ibragimov &
		Abdullabekov 1974b
Kuril-Kamchatka zone	140 ± 60, 13.6, 6.2	Fedotov 1962,
		Fedotov et al. 1976
Baykal	13.6	Lamakin 1966
Middle Asia		
(a) Fergana Depression	40, 20	Ibragimov 1970
(b) Tashkent, Kyzyl-Kum	40	Ulomov 1971, 1974
(c) Western Tien Shan	40	Ibragimov &
		Abdullabekov 1974a
(d) Pamir, South Tien Shan	260, 100–110	Nikonov 1977
(e) Chuysk Depression	15–20	Green et al. 1976,
northern Tien Shan		1978, 1980
(f) Kirghizstan	88, 22, 11, 5.03, 5.56	Kim & Aralbayev 1973
(g) Kopet-Dag	11	Abdullabekov 1979
Baltic shield	1 day, 1 year	Panasenko 1974
Japan		
(a) Japan and other	21 min, 42 min, 1 day	Aki 1968, Aki 1956
regions of the Earth	14.8 days, 29.6 days	
	6 months, 1 year, 11, 100,	
	200, 240, 284 yrs	
(b) Kamakura	69 ± 13	Kawasumi 1970
(c) Tokyo, Kyoto	36, 38	Usami & Hisamoto,
		1970, 1971
(d) Hokkaido	100	Utsu 1972a, b
(e) Central Japan	117 ± 35	Rikitaki 1979
(f) Murato	110–150	Magnitskiy 1965
Cntl America, Chile		
(Conception)	100–150	Magnitskiy 1965
China	about 1500	Mei-Shi-Yun 1960
Balkan Peninsula	35, 17–18, 11–12	Vadkovskiy
		et al. 1978
Europe	30	Karnik 1969
Indo-Australian/		
Eurasian shields	34–38	Mogi 1976

1.5 SLOW REGIONAL VARIATION OF GEOMAGNETIC FIELD

Slow regional variations incorporate anomalous changes in the magnetic field associated with processes in the Earth's crust and having typical times of 10–30 years. This variation was investigated by Orlov (1958, 1959), Barsukov et al. (1968), Abdullabekov & Golovkov (1974), Golovkov, Ivanov et al. (1977), Shapiro (1976, 1981, 1983), Rivin (1977, 1980, 1983), Skovorodkin & Bezuglaya (1980) and others.

In order to explore the slow regional variation of the magnetic field of the Earth one analysed the results measured at secular points and the mean yearly data obtained from observatory networks. The variation of the field from an epoch to epoch over the Odessa-Alma Ata section was analysed as shown in Figure 6. The drawing indicates that regional properties of the secular behaviour are accompanied by local anomalies in some segments. A closer examination has revealed that the segments with the local anomalies could be attributed to the areas for which the epoch of secular behaviour coincided with the time of seismic activity.

The relationship between the secular anomalies and the seismic activity has been analysed. The seismostatistical analysis was applied to catalogues of violent earthquakes in the Crimea, the Caucasus, Kopet-Dag, and Middle Asia (Ibragimov & Abdullabekov 1974a, b, Abdullabekov & Berdaliyev 1976). It has been found that seismicity appears in each region in a specific manner. As a rule, earthquakes are repeated with in a certain cycle. Each cycle consists of active and passive semi-cycles having the duration of 10–15 years. The secular anomalies have been observed at the places where the epoch of the anomalous secular change in the geomagnetic field coincided with the semi-cycle of seismic activity (Abdullabekov & Berdaliyev 1976).

Above 200 magnetic observatories are presently active all over the world, of which one third is located in areas with high seismicity, volcanism and intensive tectonics. Therefore the research on changes in the geomagnetic field associated with crustal processes is very effective if it employs the data from the observatories.

The secular change in the geomagnetic field is caused by various processes of internal and external origin, associated with the following features:

1. Processes in the Earth's core;
2. External processes in the Earth's magnetosphere;
3. Changes in the Earth's crust (independent or active, induced or passive).

The global and large regional changes in the geomagnetic field are brought about by processes in the Earth's core and external sources, while the local changes are caused by different processes in the Earth's crust.

The passive changes in the magnetic field are induced by external fields in individual segments of the Earth's crust through electric currents, and therefore are tightly correlated with external processes.

In order to determine the nature of the peculiar properties in the variation

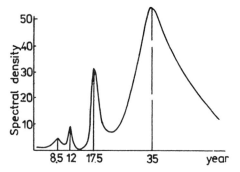

Figure 5.   Cyclicity of earthquakes in the Balkan region after Vadkovskiy et al. (1978).

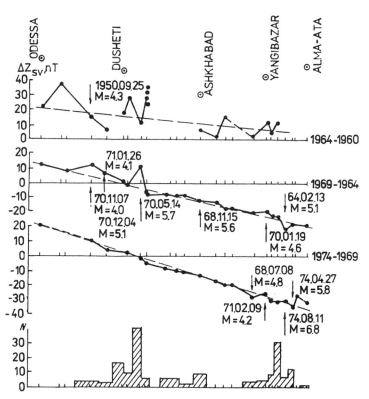

Figure 6. Variation of Z-component of secular change in magnetic field over the Odessa-Alma Ata section during the years 1964–1969 and 1969–1974. The errors identify the earthquakes occurring within the radius of 40 km from the indicated points. The lower part of the drawing encompasses the total amount of the earthquakes which occurred within the radius of 100 km from the respective points.

of the field beyond the present concepts on the sources of the variation inside the Earth and the ionosphere, the secular change in the geomagnetic field was analysed by the data of the world observatories from their beginning until the recent time (Abdullabekov & Golovkov 1974, Abdullabekov & Maksudov 1975, Abdullabekov & Berdaliyev 1976). It is known that the temporal spectrum of the secular change contains phenomena with various typical times: hundreds of years and more, about 60, 22 and 11 years and even less (Golovkov & Kolomiytseva 1976, Yanovskiy 1963, 1978, and others). The sources of 60-year changes lie in the Earth's core, while shorter periods in the secular change seem to be associated with processes in both the core and beyond it; their amplitude being small (first tens of nanoteslas) and they themselves being periodic (Rivin 1977, 1980, Golovkov & Rivin 1976, and others). In some observatories one obtained sharp single features of the secular change having periods about 10–30 years and an amplitude of 100–300 nT (Abdullabekov & Golovkov 1974). The nature of the latter components is linked to the processes in the Earth's core as well as to local and regional geodynamical processes. Detailed spatial and temporal analysis of these variations will probably make it possible to distinguish local changes from the global ones.

Hence the changes in the geomagnetic field appear in a very wide temporal spectrum, from fractions of seconds to hundreds and thousands of years. The amplitudes of these variations can be compared by magnitude with local ones, or are even much higher. The characteristic times of the latter are closer. Therefore the discriminaton of local properties in the secular change from a series of observations at a single station is impossible if one uses different frequency and time filters.

Spatial distributions only can be employed in the separation of field changes associated with different processes in the Earth's crust. Indeed, the sources at a large depth (greater than half-radius of the Earth) generate large regional changes.

Thus, the identification of local changes in the geomagnetic field is possible only by the use of different spatial filters.

In the identification of characteristic properties of the secular change, the series of X, Y, and Z-components of the geomagnetic field or $\dot{x}, \dot{y}, \dot{Z}$−components of the secular change, smoothed out through 11-year shift averaging, have been mutually compared for a group of close obsevatories (Fig.7). Golovkov & Kolomiytseva (1970) have shown that 60-year changes in the geomagnetic field of Eurasia in the years 1900 to 1950 were caused by the source located at 27°N, 41°E, and have a regular course.

The temporal behaviour of the secular change can be approximated by the following exponential function:

$$\dot{F}(t) = -A \ t \ exp[-(\frac{2t}{T})^2] \tag{1.4}$$

in which

  A = numerical parameter,
  T = quasihalf-period (of 32 years).

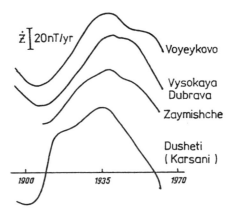

Figure 7. Secular change in $\dot{Z}$-component of magnetic field at observatories of Voyevkovo (Leningrad), Vysokaya Dubrava (Sverdlovsk), Zaymishche (Kazan) and Dusheti (Tbilisi).

The secular change curves shown in Figure 8 also obey this law but, by the data from Dusheti (Karsani) for the years 1908–1925, they were characterized by the maximum deviation from the normal behaviour reaching 40 nT per year. This property has been compared versus seismicity (Abdullabekov & Berdaliyev 1976).

Violent earthquakes in the Caucasus are attributed to particular seismotectonic zones, and appear in cycles having the period about 35 years (Ibragimov & Abdullabekov 1974). The results of the comparison of the secular change versus seismicity have shown that the anomalous behaviour of the magnetic field at Karsani in the years 1911–1928 coincided with the period of seismic activity in the Great Caucasus. Analysis of the catalogue and the chart of violent earthquakes indicates that numerous perceptible earthquakes occurred at that time, including the violent earthquakes at Gori in 1920, with $M = 6.5$, the epicentre of which was 60 km from the observatory. It is characteristic that the epicentres of earthquakes and the Karsani observatory have been situated in a unique seismogenic zone and moreover the isoseists of that earthquake were extended along the seismogenic zone in the parallel direction (Kirrilova, Lustykh, Rostvorova et al. 1960).

There arises a question why the anomalous change in the geomagnetic field did not occur during the subsequent active period. It appears that it was within the seismogenic zone before the observatory was moved from Karsani to Dusheti.

Figure 8 illustrates the variation of mean yearly Z-components in the years 1955–1975 obtained at the observatories over the section Wingst-Grocka. The clear difference of the secular change at the observatories Vienna-Kobelnz is obvious. At the observatories Vienna-Kobelnz, Regensburg and Fürstenfeldbruck during the above period one observed quite another secular change, compared with the other observatories of Central Europe. The Z-component at these observatories, beginning from 1955 was falling down slowly with regard to the normal change. At the adjacent observatories of Prugonice, Hurbanovo, and Tihany located some 150–

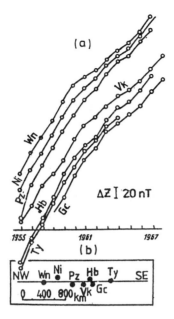

Figure 8. Variation of yearly Z-components of magnetic field along the Wingst-Grocka profile (a) and layout of geomagnetic observatories (b): W_n = Wingst; N_i = Nimeg; P_r = Prugonice; N_K = Vienna-Kobenzl; H_B = Hurbanovo; T_y = Tihany, G_c = Grocka.

250 km from the observatory Vienna-Kobenzl, the peculiarity was not perceptible.

The observatories in which peculiarities of the secular change were observed are located along one active tectonic structure, that is the Pre-Alpine fault, parallel to the Alpine meganticlinorium. There are numerous Alpine faults which cut through the primary geological structure (Fig.9).

Hence the above property of the secular change is linear. The stretch of the anomalous structure of the secular change coincides with the direction of the primary geological structure. The linear dimension of the identified anomaly is about 300–400 km, its density counted in tens of nanoteslas, and the anomaly has a linear elongated shape.

The spatial and temporal characteristics of that anomaly were later studied in detail by Kuznetsova et al. (1975) and Maksimchuk (1983). They analysed the data from the network of observatories in the Carpathian-Balkan region. The principal conclusion of Abdullabekov & Golovkov (1974) about the presence of a regional anomaly with a characteristic dimension of first hundreds of kilometres and the density of tens of nanoteslas has been confirmed.

Regional anomalous field changes with similar spatial and temporal parameters have recently been discovered in the data of repeated route and aeral surveying in the Urals, Alma Ata, Crimea and other test areas. Of particular value is the result obtained by Shapiro et al. (1983) in the Ural area. A fairly dense grid of stations

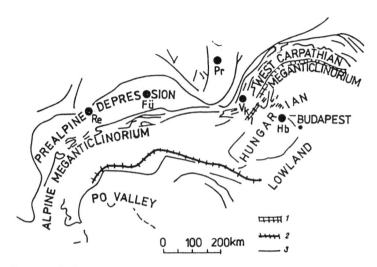

Figure 9. Layout of observatories Regensburg (Re), Fürstenbruck (Fü), Vienna-Kobenzl (VK) on the tectonic chart of Central Europe (copied from the tectonic chart of Euroasia compiled by Yanshin et al. 1966): 1–3 = discontinuities: 1 = ridge type; 2 = major deep-reaching faults; 3 = undifferentiated faults.

was deployed in an area of 400 km², where regular surveying of high accuracy was conducted with the aid of TMP proton magnetometer designed at the Institute for Geophysics of the Ural Division of the Soviet Academy of Sciences; its sensitivity was 0.1–0.2 nT.

From the data of aeromagnetic surveying in the years 1959–1979, repeated ground surveying in the years 1968–1980, and variational observations at Arti and Vysokaya Dubrava it has been found that the anomalous field within the Manchazhskaya anomaly increases in 10–15 years at a rate of 5 nT per year (Shapiro 1973). Shapiro believes that the anomalous changes are caused by the variation of the residual magnetism of rock due to excessive stresses.

Hence the analysis of mean yearly data of the world network of magnetic observatories, together with a long series of measurements in test areas and at points of secular changes, has discriminated a slow variation of the magnetic field with an intensity of 50–70 nT and a characteristic time of 15–20 years, having the linear dimensions of the first hundreds of kilometres. This type of variation has been explained in terms of tectonic processes in the active areas of the Earth's crust and phases of seismic activity in seismotectonic regions.

CHAPTER 2

Prototype measurements of magnetic field at variable loading

2.1 ANOMALOUS ANTHROPOGENIC CHANGES IN MAGNETIC FIELD

Many cases of the variability of the Earth's magnetic field due to various processes in the Earth's crust are already well known. The identified modes of variation are usually explained in terms of excessive stresses prior to earthquakes and thermal, chemical, electric or other factors. However, it is only qualitative interpretation that is usually provided. This is due to our limited knowledge and the complexity of the processes occurring in profound strata. An unambiguous interpretation, aside from full description of spatial and temporal variation of the field, also requires a linkage of the field changes to different factors such as pressure, temperature, physical properties of rock etc. The information needed for those parameters can be inferred from measurements of magnetic field under controlled loading conditions caused by different technological impacts on the Earth's crust, for instance impoundment of large reservoirs, underground gas storage, or artificial explosions.

Experimental studies on the artificial effects caused by explosions have been carried out by Shapiro & Undzenkov (1967), Barsukov & Skovorodkin (1969), Kozlov et al. (1974), Akopyan et al. (1973, 1976) and others.

Shapiro & Undzenkov (1967) identified the magnetic effects resulting from explosions in magnetites. The variation of the field was measured at the three-component magneto-variational station of Bobrova, with a sensitivity of 0.13 nT/mm. The strength of the discriminated effect was counted in some nanoteslas. Barsukov & Skovorodkin (1969) used a similar method at Medeo (Kazakhstan) to obtain field variations with an intensity of 0.7 nT in granites with a residual magnetism of 10–300 $\times 10^{-6}$ [cgs]. The effect was exposed upon comparison of the measured modulus of the total vector of magnetic field T at a number of points prior to and after the explosion. Some days after the explosion the magnetic field returned to the original level.

Extensive studies on the variation of the magnetic field during explosions were conducted in the years 1973–1974 by Kozlov et al. (1974). Unlike the earlier investigations, the measurements by the above researchers employed a quantum magnetometer with a high discrimination in the time domain and a high sensitivity.

Kozlov et al. managed to identify three types of magnetic effects, i.e. fast alternating (reversible), irreversible and relaxation ones.

The fast reversible variation has been associated with reversible changes in magnetic properties at the time of passage of elastic waves. Their amplitude varies from hundredths and tenths of nanotesla to tens and more nT. The irreversible changes, with an intensity from tenths to first tens of nanoteslas were controlled by irreversible changes in rock magnetism in the disturbance zone. The field changes of the relaxation type have been linked to the relaxation of elastic stresses in the zone of explosion. The characteristic time varied from seconds to several hours and more, while the amplitude ranged from a few nT to 10 nT.

Hence the experiments on the local variation of magnetic field due to explosions in rocks conducted in various regions were carried out in different geological-tectonic, physico-mechanical and other natural and artificial conditions (weakly, medium and strongly magnetic rock at distances of 50 to 2000 m from the centre of explosion).

The strength of the explosions varied from 100 to a few thousand kilograms of explosives. As a result, a unique relationship has been obtained between the arising mechanical stresses and the changes in the magnetic field (Skovorodkin 1980). It has also been established that the intensity of the effect depends on the distance from the point of explosion and the rock magnetism.

Many geodynamical test areas have recently been employed in investigations of local changes in the magnetic field caused by water impoundment conditions in reservoirs (Davis & Stacey 1972, Abdullabekov et al. 1976, 1977, 1979, Berdaliyev et al. 1981, Berdaliyev 1981, Akopyan et al. 1976, Abramov et al. 1981, Oganesyan et al. 1979).

The first investigations were carried out by Davis and Stacey (1972) in south-eastern Australia in the region of Talbingo Reservoir. The reservoir was about 3 km long and 1 km wide, and the dam was 150 m high. The base station was located 35 km away. The measuring points arranged on both sides of the reservoir, along its banks, have permitted identification of field fall from 2 to 8 nT.

2.2 CHARVAK RESERVOIR EXPERIMENTS

Charvak Reservoir is situated 70 km north-east of Tashkent. The dam is 168 m high, 10–12 km long and 2–5 km wide. The volume of impounded water is about 2×10^9 m^3. The reservoir stores the water conveyed by Rivers Chatkal, Pokem and Koksu. The impoundment took four months (from February to May) and the volume of water increased from 2×10^8 to 1×10^9 m^3.

From May to early July the reservoir is filled up. From July to September or October the water is used for irrigation (discharge phase) while in the remaining part of the year (September to January) the reservoir remains empty. The impoundment began in 1973, and the water conditions have varied cyclically since.

The reservoir is located in a region with a complex tectonic structure and high seismicity. The reservoir zone and its neighbourhood have suffered from numerous fine and a few violent earthquakes: Pskem (1937, with M = 6.5), Brichmulla (1959; M = 5.7), Tavaksay (1977, M = 5.6) and others (Shebalin & Kondorskiy 1977). The recent chart of seismic regions places the Charvak area in the eight-degree zone (Abdullabekov et al. 1980b). The anomalous field in the test area is about ± 200 nT.

The Charvak area possesses the necessary natural conditions favouring experiments on the effect of varying pressure on the magnetic field of the Earth such as variable cyclic conditions of water level and volume variation, low background of natural and artificial noise, presence of rock with optimum magnetism, and known piezometric properties of rock (from laboratory measurements) (Maksudov et al. 1971, Maksudov 1972).

The measurements were carried out by repeated route surveying and stationary measurements (Fig.10). The points of route surveying established in the years 1973–1975 were situated from 1–2 to 5–6 km from the reservoir banks, and generated the enclosed polygon Gazalkent-Chimgan-Yusupkhana-Aurakhmat-Brichmulla-Bogustan-Nanay-Sidzhak-Charvak-Gazalkent. The observations were conducted every 1–3 months by the use of proton magnetometers (PM-5, TMP-1, Zh-816 and PMP). The measurements in the years 1973–1978 were related to the base line T of the variometer of Yangibazar Observatory, which is located 40–50 km south-west of the test area. The method of synchronous measurements has been used since 1978.

The stationary measurements were accomplished with the aid of an APM proton magnetometer (Berdaliyev et al. 1980), round-the-clock every twenty minutes (Fig.10).

The mean-square accuracy of the measurements in repeated route surveying was about 1 nT, while that for the stationary measurements was 0.6 nT. The measuring error in the route observations was assessed by special 2–3 cycles of inspection measurements by the same team or by one-cycle measurements by several teams. The measurements were carried out at different times of the day, with different instrumentation, and by various staff.

The mean-square accuracy of the stationary measurements was assessed by results of the synchronous measurements of the magnetic field conducted during some days at Yangibazar and at Charvak Reservoir, with a time increment of 20 minutes:

$$\sigma = \sqrt{\frac{\sum \Delta T^2}{2n}}. \tag{2.1}$$

Since the route measurements were conducted with different proton magnetometers in consecutive years, particular attention has been paid to the assessment of the accuracy of instruments. Voluminous methodological work has been done. The magnetometers were repeatedly calibrated against the central magnetic observatory (TsMO) 'Krasnaya Pakhra' and Yangibazar Observatory. The accuracy

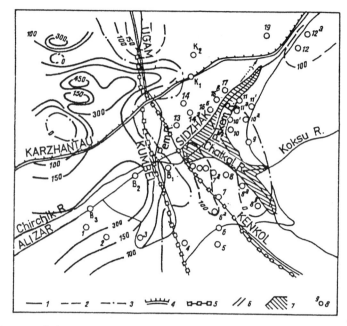

Figure 10. Layout of observation points in Charvak test area: (1–3) isolines of anomalous magnetic field ΔT_a in nanoteslas; (1) positive; (2) negative; (3) neutral; (4) overfault and fault; (5) late Alpine faults; (6) contemporary faults; (7) Charvak Reservoir contour; (8) points of repeated measurements of magnetic field.

was also evaluated by multiple synchronous measurements at the same place.

Results of repeated route observations. Beginning from December 1973 the following repeated observations were carried out in the test area of Charvak Reservoir: five in 1974, eight in 1975, six in 1976, three in 1977, seven in 1978, and six in 1979. From 1980 onward the observations were rarer, i.e. two to four times per year (Abdullabekov et al. 1976, 1977, 1978, 1979, Berdaliyev et al. 1981).

Characteristic field changes, correlated inversely with the reservoir conditions, have been established at 35 points of the test area. Moreover, some individual properties depending on the position (distance from bank and geological-tectonic factors) have also been identified. The anomalous deviation at the points associated with the impoundment conditions varied from 1–2 to 7–8 nT. The relationship of the anomalous field variations and water level in the reservoir was clearly repeated from year to year (Fig.11).

The intensity of the anomalous change decreases with distance from reservoir bank (Fig.12). The most dramatic changes are attributed to the northwestern part of the reservoir. It appears that the anomalous variation also depends on the

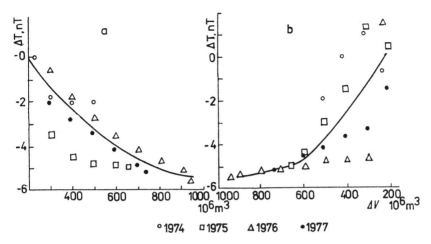

Figure 11. Dependence of anomalous magnetic field change on water volume in reservoir upon (a) filling and (b) emptying.

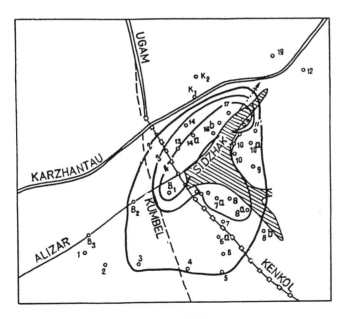

Figure 12. Distribution of anomalous magnetic field changes over area upon impoundment of Charvak Reservoir.

geological-tectonic conditions of the test area. The role of the reservoir is most pronounced in fault zones.

Results of stationary observations. In order to investigate more thoroughly the relationship between the magnetic field changes and the variable pressure conditions, stationary observations at the reservoir were initiated in 1978. The variation was recorded round-the-clock every twenty minutes and simultaneously with Yangibazar Observatory (Abdullabekov et al. 1979, Berdaliyev et al. 1981).

The results of the stationary measurements in 1978 and subsequent years coincide fully with the results of the repeated route surveying. However, the mean daily difference varies in a more complex manner than the reservoir condition (Fig.13). Aside from the slow variation one also observes a clear quasi-periodic variation with the characteristic time of 15–30 days. These insignificant deviations of mean daily differences from mean decade values are so far difficult to explain. It is possible that the high frequency component of the curve is associated with a sophisticated reaction of rock to the additional loading or that it is simply a difference of field at two points located in different blocks.

In the present experiment it is important to identify the slow (low-frequency) variation of the field. When the water volume increases by 1×10^6 m^3, the magnitude of the magnetic field decreases by 5–7 nT. This regularity remains practically unchanged at all points (Fig.13a-b).

Hence we have established experimentally a relationship between the change in magnetic field versus the water volume in reservoir. The obtained result cannot be explained in a unique way. The variation of the magnetic field may also be controlled by additional loading (piezomagnetic effect) or higher water saturation of rock (electrokinetic effect).

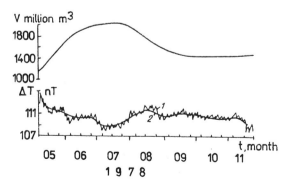

Figures 13. Results of stationary measurements of magnetic field at Charvak in 1978: (1) mean daily; (2) mean decade value.

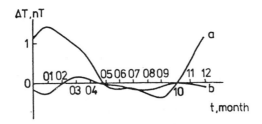

ΔT,nT

t,month

Figure 14. Variation of magnetic field above 'wet' (a) and 'dry' (b) rock.

2.3 RESULTS OF EXPERIMENTS IN REGIONS OF SPRINGS AND UNDER-
GROUND GAS STORAGE: ELECTROKINETIC EFFECT

We have discussed the variation of the magnetic field associated with a direct effect of stresses on magnetic properties of rock in laboratory and under field conditions (the piezoeffect). This variation can however be also of electokinetic nature. We will dwell on the results of experiments conducted about reservoirs and above a dome of a natural undeground gas storage.

Investigation of the effect of springs on magnetic field. Eight years of investigations were devoted to the above objective at all points of the Tashkent test area; the results have been processed statistically. The points have been categorized in two groups: points close to irrigation gutters (so-called 'aryks'), small brooks and rivers up to 3 m wide in the first group, and points located far away from water resources in the second group. The analysis has shown that water resources might affect the magnetic field (Fig.14). The curve of yearly variation of ΔT on humid rock has a clearly pronounced cyclic yearly course (Fig.14a). On the other hand, the curve for the remaining points on dry soil is practically constant over the year (Fig.14b).

Under conditions of the Tashkent region, the upper part of the cross-section (daylight) consists of sedimentary rock with high porosity and permeability (sandy loam, loam clay, loess).

In the case of impoundment the rock soon becomes saturated with water and its electrical conductivity increases sharply, which should be visible in the mean daily variation (S_q). However, as shown by investigations within the testing area of the gas storage and Charvak Reservoir, the variation of S_q was identical with that on the territory of Yangibazar Observatory. Hence the role of electrical conductivity of rock at reservoirs in the generation of anomalous changes is negligible.

Accordingly, the nature of anomalous changes in the geomagnetic field is explained by two factors, that is excess loading and the electrokinetic effect. The depth of reservoir and the relative variation of water level within the test area do not exceed metres. Hence the excess pressure is below 100 kPa. If the magnetism of the Quaternary deposits is 10^{-8}–10^{-6} [cgs] the field variation due to the additional loading of the order of 10–30 kPa amounts to hundredths of nanotesla. Taking into account the above circumstances and considering the velocity of water flow in reservoir one can assume that the anomalies identified have been caused by the electrokinetic effect (Abdullabekov & Sultanbekov 1976, 1978).

The variation of the magnetic field caused by the electrokinetic phenomena was also identified under laboratory conditions (Migunov & Kokorev 1977). The generation mechanism of magnetic field by electrokinetic factors has been described by many authors (Migunov & Kokorev 1977, Rakhmatulin & Sultanbekov 1976). The electrokinetic phenomena are pronounced more strongly with decreasing cross-section of voids and increasing hydrodynamical pressure. In our case the rock (sandy loam, loam clay and loess) has high porosity, which seems to cause the insignificant anomalous variation (< 2 nT).

Investigation of geomagnetic field variation due to changes in gas conditions of gas storage. Detailed magnetometric observations were conducted in the region of an underground gas storage. Natural gas was pumped at 9–9.5 MPa into an anticlinal aquifer at a depth of 600–650 m. The anticlinal structure had a size of 6 × 1.5 km. Eleven points were deployed in that area, and repeated observations were conducted 3–5 times a year (Abdullabekov 1972, Abdullabekov & Golovkov 1974, Abdullabekov & Sultanbekov 1976, 1978). The gas storage consisted of a collecting layer at a depth of 520 m at arc and 720 m on sides of a brachyanticlinal structure 20–40 km thick. Paleozoic (1300 m), Mesozoic (300 m) and Cenozoic deposits exist in the stratigraphic structure of the rise. The structure is asymmetric with a steep northern side and mild southern one, having a length about 6 km and a width of 2 km. The collecting layer is overlain by 80 m of clay. The pressure is 9–5 MPa. The natural gas is pumped into that storage, and the stratum pressure of the collecting layer is 6 MPa.

Figure 15 illustrates the mean yearly variation of the magnetic field (ΔT) and the gas pressure in the gas dome (P) from 1968 to 1975. A comparison of the graphs shows that the variation of ΔT and P are correlated: decreasing pressure from November to April corresponds to increasing ΔT, while increasing pressure from May to October is accompanied by decreasing ΔT, with the maximum pressure in October standing for the minimum value of ΔT.

Barsukov & Mutaliyev (1972) conducted magnetometric and electrical measurements by the method of dipole electric probing in the gas storage region. The western dome of the investigated structure exhibited apparent specific resistance by 15–16% higher due to penetration of gas into voids at a pressure of 1–1.2 MPa. On the other hand, a lower apparent electrical resistivity (by 7–8%), was observed at the eastern dome, which corresponds to a higher pressure of water in the layer (by 200–300 kPa). During the period of intensive evacuation of gas and respective fall in the rock pressure one observed an increase in the apparent electrical resistivity (by 7–8%).

Let us discuss the possible nature of anomalous magnetic field changes above the

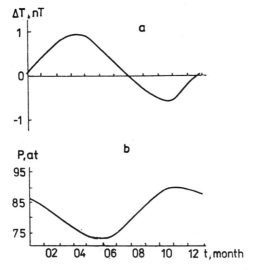

Figure 15. Anomalous field changes associated with pumping of gas into and out of underground storage. (a) variation of magnetic field; (b) variation of pressure in gas domes of storage.

gas storage. The rocks in the rise have low magnetism ($1 \times 10^6 - 1 \times 10^{-5}$ [cgs]). The maximum excess pressure does not exceed 2.5–3 MPa, on the average. Hence the piezomagnetic effect is not negligible as it is counted in tenths of nanotesla. The temperature and chemical composition of the gas remains practically constant during the procedures of pumping in and out. Thus, the thermal chemical processes cannot bring about substantial anomalies in the magnetic field changes.

Further to the above discussion let us consider the variation of the magnetic field upon relative motion of the liquid and solid phases. If one assumes that the electrokinetic phenomenon controls the anomalous changes in the magnetic field then the analytical relationship in the magnetic effect and the primary parameters of the gas dome should read (Abdullabekov & Sultanbekov 1978):

$$\frac{H_2}{H_1} = C \frac{W_1}{W_1 - \frac{BT_2}{P_2} M_2} \tag{2.2}$$

in which
H_1 and H_2 = strength of magnetic field prior to and after gas pumping, respectively,
W_1 = volume of electrokinetically active water-permeable stratum prior to gas pumping,
M_2, P_2, T_2 = mass, pressure and absolute temperature of gas, respectively,
C = coefficient of electrokinetic activity of collecting layer, depending on volume of gas dome,
B = gas constant.

All parameters of the gas dome but P_2 are constant. Accordingly Equation 2.2 can read:

$$\frac{H_2}{H_1} = C(1 + \frac{1}{\frac{W_1}{A}P_2 - 1})$$ (2.3)

in which
$$A = BT_2M_2.$$

The graphical representation of Equation 2.3 is a hyperbola with the asymptotes $\frac{H_2}{H_1} = C$ and $P_2 = \frac{A}{W_1}$. The relationship between the computed magnetic field and the pressure in the gas dome is inversely proportional. The maximum values of pressure correspond to minimum $\frac{H_2}{H_1}$, in agreement with experimental data (Fig.16).

From the above discussion it can be concluded that not all anomalous changes in the geomagnetic field prior to earthquake can be linked to tectonic stresses. The variation of the geomagnetic field can also be caused by electrokinetic phenomena under some particular conditions.

The nature of the anomalous changes in the geomagnetic field at Charvak Reservoir is much more complex than the effect above the gas storage and at reservoirs. This is so because the gas storage and shallow reservoirs have very local dimensions and their rocks can be considered homogeneous in the magnetic sense. Charvak Reservoir occupies a large area. The magnetic properties of rock over there are diversified in wide ranges. This seems to cause the differentiation of the intensity of anomalous changes at different points (from unity to the first tens of nanotesla). At the same time, the tendency of changes is identical at each point. The increasing volume of water corresponds to lower magnitude of field, and vice versa. The course of field variation at Charvak Reservoir versus water volume (water level) is identical with that at the gas storage and in wet rock. However, the intensity of the anomaly is here 4–5 times greater than in the former case. This seems to be associated with the large variation of water level and considerable magnetic properties of the rock. The tectonomagnetic effect should be given specific attention in investigations at Charvak Reservoir (Abdullabekov et al. 1977, 1979). However, experiments have shown that the anomalous changes in the region of the reservoir, as typical of reservoirs and the gas storage, cannot be explained uniquely by the tectonomagnetic effect. A certain role in the anomalous changes is also played by the electrokinetic effect.

The electrokinetic effect has been known for long time in physics (Anonymous 1956), while in geophysics it has been given attention quite recently (Rakhmatulin & Sultanbekov 1976, Abdullabekov & Sultanbekov 1976, 1978, Kormiltsev & Shapiro 1979, Mizutani & Ishido, 1976, Mizutani et al. 1976, Skovorodkin & Bezuglaya 1980). The papers cited show that the magnitude of the effect due to small reservoirs and groundwater is counted in first units of nanotesla. The anomalies can also be very high (up to 200 nT) due to dilatancy of rock at focuses of incipient earthquakes.

Hence particular role should be played by the electrokinetic effect in experiments on Uzbek test areas where sedimentary rock, favouring seepage of water, is widespread. The electrokinetic effect appears principally as an interfering factor in seismomagnetic investigations, but may also be a useful signal in some cases. For instance, a dramatic change in the groundwater conditions (variation of water level in wells) has been observed before many violent earthquakes.

CHAPTER 3

Tectonogeological and geophysical features of Middle Asia and selection of geodynamical test areas

3.1 TECTONOGEOLOGICAL CONDITIONS

From the geologico-geomorphological point of view, Middle Asia and southeastern Kazakhstan can be divided into Pamir-Tien Shan highlands, Turan shield and Kopetdag. The area of investigations is bounded by central Kazakhstan in the north, Tarim Platform in the east, the folded systems of Kunlun, Karakorum, the Himalayas and the Tibet shield in the southeast, the Indian Platform and the Hindu Kush folded system in the south, and the Russian Platform and the West Siberian Plain in the west.

The geological structure, tectonics, seismicity, geophysical fields and profound structure of the Earth's crust of Middle Asia, along with the surrounding large folded systems and rigid massifs, make up a very specific system.

One notes the following four primary geological properties (Kharambayev 1977):

1. The presence of a distinct sublatitudinal tectonical zonality pronounced in the gradual change from the north to the south of the geosynclinal-folded structures of Caledonian (the northern Tien Shan), Hercynian (the central and southern Tien Shan and the northern Pamir) and Alpine (the central and south-eastern Pamir and Kopedag) origin. The Tien Shan arcs are convex to the south while those of the Pamir are arched to the north;

2. Lateral and vertical inhomogeneity of the Earth's upper crust;

3. The presence of 'translucent' structures;

4. Periodically repeating sequence of the processes of sedimentation, magmatism, metamorphism and tectogenesis, which bring about the formation of structural levels, horizons, etc. The Alpine, Hercynian, Caledonian and Baykal levels are distinguished in the downward direction.

Some other properties of the territory include the presence of the highest mountain systems, the expansion of the crust thickness from 40 km in the north to 70 km in the south, frequent seismic activities with a magnitude of 7.5–8 and more, etc. The territory of Middle Asia has been an arena of the Soviet (all-union) and international investigations. Soviet-American seismological investigations were conducted in one of the geodynamical areas (Garm). Since 1971 a number of countries have participated in the International Pamir-Himalayan Project (Pamir-Himalayas 1982).

The highest mountainous foldings of the Pamir and the Tien Shan were generated in recent times (the former at the end of Eocene, and the latter in Oligocene) as a replacement of platforms.

Tectonic regions. A fairly large selection of tectonogeological regionalization schemes have been proposed for the territory of Middle Asia. The earliest and well substantiated is the scheme proposed by Nalivkin (1926). It has not been changed substantially for a long time, and many researchers (Popov 1938, Abdulayev 1960, Ulomov 1974, Khamrabayev et al. 1977, and others) still support this concept. The scheme identifies the northern, central and southern Tien Shan, the northern, central and southern Pamir, Kopetdag and other regions (Fig.16).

The northern Tien Shan zone consists of Caledonian folding structures subdivided into Karatau-Talas and Kirgiz-Terskey subzones. The Muyunkum-Narat massif adjoins in the north, while the Nikolayev line is the boundary in the south (Knauf & Korolev 1974, Yerzhanov et al. 1982). The Caledonian level of the zone consists of Cambrian, Ordovician and Silurian rocks of total thickness of 5–7 km. The rocks are strongly metamorphosed sedimentary-volcanic and terrigeneous-volcanic and orogeneous formations (molasses). The primary phase of the folding belongs to the middle Ordovician. The Terskey, Kokdzhat and Ichkelitau anticlynoriums and the north Kirgizian, Susamir and Talas-Karatau synclinoriums are distinguished within the northern Tien Shan.

The zone of the central Tien Shan, encompassing the Kurimin, Chatkal, Naryn and Great Karatau subzones consists of Precambrian, Cambrian, Ordovician, Devonian and Carboniferous rocks, 6–7 km thick in total. The primary phase of the folding consists of middle Carboniferous rocks.

The zone of southern Tien Shan is divided into the subzones of Aumin-Turkestan-Zarafshan, Kokshaal-Yassin, Baysun and others. The northern boundary passes over the deep-reaching faults of north Nuratin–south Fergana and Atbashi-Inylchek. The zones consist of upper Proterozoic, lower Paleozoic (Prehercynian formations) and upper Paleozoic (Hercynian deposits). The primary phase of the folding was generated in middle Carboniferous.

The north Pamir zone is incorporated in the Kunlun system (Barkhatov 1971), and is bounded from the north by the deep-reaching fault of north Pamir, while the fault zone of Akbaytal adjoins in the south. It consists of Precambrian and Paleozoic rock of sedimentary-metamorphic and volcanogenic-terrigeneous composition. The primary phase of the folding was generated in Permian.

The central Pamir is bounded by the Akbaytal fault zone in the north, while the deep fault of Rushan-Pshart is the southern border. The zone includes Vanch-Yazgulem, Muzkol-Sarykol and Rushan Pshart subzones. Precambrian, Paleozoic, Mesozoic and Cainozoic deposits participate in the structure of the zone.

The zone of southern Pamir consists of south-west and south-east subzones, and is bounded from the south by the deep-reaching fault of the southern Pamir Knot.

Recent movements. As noted, the Tien Shan territory and the Turan Plain have recently remained in the platform stage of growth. The territory of Pamir has risen intensively since Eocene, while the Tien Shan has grown since Oligocene. The total span of the recent tectonic movements (rise of ridges and fall of depressions) over 25–30 million years reached 11–14 km. The intensity of the recent movements differs in time and space (Table 2). Segments of rise and fall alternate (Nikolayev 1962, Ibragimov 1968, Rantsman 1979, Gorshkov 1984 and others).

The Turan shield territory occupies a wide area of deserts, semideserts and piedmont oases of Middle Asia. In tectonic categories it is an epiPaleozoic platform.

Information about the recent movement of the Turan shield is contained in works by Yanshin (1948), Petrushevskiy (1955), Borisov (1958), Nikolayev (1962), Mavlanov et al. (1969), Ulomov (1964), Ibragimov et al. (1973), Yur'yev (1967), Sitdikov (1976) and other studies.

The total span of the recent movements varies from tens to hundreds of metres, or up to 1000 m and more in some cases. For instance, the Tashauz depression of the Paleozoic base in the western part, to the south of Urgench, has measured some kilometres. The maximum value at places has reached 3–5 km (Nikolayev 1962). The span of the recent movements is insignificant in the remaining part of the Turan shield (north-west and south).

The magnitude of recent movements is also counted in hundreds of metres in the central Kyzyl-Kum. At the highest points the recent movements reach 1 km at places (Bukantau and Auminzatau) and 700–900 m (Kuldzhuktau). There are some small segments of depressions with an amplitude up to 200–300 m (Ibragimov et al. 1973).

Hence, compared with the orogenic part of the Turan Plain, the magnitude of the recent movements is insignificant. Some regularities are observed. The topography and growth of structures are controlled by the recent movements. The structures are elongated in the north-west direction and are well pronounced on charts of recent movements and anomalous magnetic and gravitational fields. At the same time, however, there are discontinuities and anomalies of the gravitational field in the south-west direction. The latter seem to be associated with the recent structures of the central Kyzyl-Kum.

The zone of intensive orogeny is contained in Paleozoic plateau structures. The mountainous topography was generated in Neogene-anthropogenic times as the result of intensive tectonic movements of the crustal segments which were earlier stagnant during a long period of time.

There is a direct functional relationship between the magnitude of neotectonic rise and the thickness of the Earth's crust: the stronger the increase in the crust thickness, the higher the rise (Gzovskiy 1961, Krestnikov 1958, and others).

Discontinuities. There are numerous discontinuities within the mountainous struc-

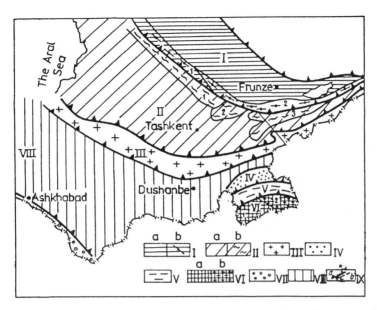

Figure 16. Tectonogeological regions of Middle Asia after Khamrabayev et al. (1977). Structural-formation zones: (I) Northern Tien Shan: (a) Muyunkum-Narat central massif, (b) Caledonian subzones (1 = Karatau-Talas, 2 = Kirgiz-Terskey); (II) Central Tien Shan: (a) Kuramin central massif, (b) Hercynian subzones (1 = Great Karatau, 2 = Chatkal, 3 = Naren); (III) Southern Tien Shan (Hercynian); (IV) northern Pamir (Hercynian); (V) central Pamir (Alpine); (VI) southern Pamir: (a) south-western subzone (central massif, 1); (b) south-eastern subzone, 2; (VII) Kopetdag (Alpine); (VIII) Karakum-Karakum Tadzhik (border massif); (IX) large fold disturbance: (a) intrazonal deep-reaching folds, (b) other deep-reaching folds.

Table 2. Amplitude of recent tectonic movements after Ibragimov (1978).

Level	Rise of ridge, m					Fall of depression, m		
	Kur	Cha	Fer	Alay	Tur	Fer	T-G	Cha
Oligocene-Lower Miocene	300	500	500	1000	1000	1000	300	500
Middle-Upper Miocene&Lower-Middle Pliocene	1500	1800	1900	2500	2500	2500	800	1700
Upper Pliocene	1600	1800	1800	3000	2800	2500	800	600
Anthropogenic	500	600	700	1000	1200	1200	400	200
Total ampl.	4000	4000	4500	6500	6500	7000	2300	3000

(Kur) Kuramin; (Cha) Chatkal; (Fer) Fergana; (Tur) Turkestan; (T-G) Taskent-Golodnostep.

tures of the Pamir-Tien Shan and the Turan Plain, differing by direction, time, depth and other properties.

A lot of attention is paid to discontinuities in almost all works on the tectono-geological and seismotectonic features. Special studies devoted to faults are very scarce. They include a contribution by Yakubov et al. (1976), in which faults are classified by age, morphology and other characteristics, and information is given on the most important parameters (depth of penetration, age characteristics, time of activation, gravity steps, etc).

In the western Uzbekistan there predominate systems of north-western and sub-lalitudinal faults, but north-eastern and submeridional features are also encountered, together with NNW and ENE in some particular cases.

North-eastern and sublatitudinal faults prevail in the central Tien Shan but north-western configurations are also encountered. A dense grid of faults is observed in the remaining territory, that is the northern Tien Shan, Pamir etc.

A variety of discontinuities, with different directions, depth, morphology and stretch appear in the tectonic systems of the Tien Shan and Pamir. However, not all faults are active at present. In the opinion of many researchers (Akhmedzhanov et al. 1971), the primary network of faults was generated in Archaeo-Proterozoic, while the faults of definite orientation and category were activated in various cycles of tectogenesis.

Ulomov (1974) distinguishes systems or zones of faults having a width from 20–30 to 70–100 km and a length of hundreds kilometres, in lieu of individual faults. They include northern Tien Shan, Naryn, Talas-Fergana, southern Tien Shan, central Kyzyl-Kum, western Tien Shan, Pamir-Alay, and central Pamir.

The northern Tien Shan fault zone stretches along the northern feet of Kirgizian Ridge and Zailiyskiy Alatau. The focuses of the strongest and catastrophic earthquakes with a magnitude up to 8 and more are attributed to this zone (earthquakes of Belovodsk 1885, Verne 1887, Chilik 1889, Kebin 1911, etc). In the opinion of many investigators (Nersesov et al. 1980 and others), a consecutive seismic activity is expected in the nearest time, as no seismic events with $M > 7$ have been observed in that zone since the strongest Kebin earthquake of 1911.

The Naryn fault zone separates Caledonian and Hercynian features of the Tien Shan, and is less active in seismic terms.

The Talas-Fergana fault zone of north-west orientation passes through the eastern Middle Asia structures of different age (Caledonian and Hercynian). The most clear-cut feature of the fault zone is the right-hand-side displacement at a distance of 70 to 200 km (Burtman 1964, and others).

The activity of the fault varies over 800 km. The north-western part is practically aseismic, while earthguakes of average energy class (M = 5–6) have been encountered in the south-eastern part. It is controversial if the focus of the 1946 Chatkal earthquake with $M \approx 7.5$ could be attributed to that zone (the focus of the earthquake is somehow shifted to the west and can be better associated with the Chatkal-Ataynak fault).

The southern Tien Shan fault zone (Gissaro-Kokshaal seismogenic zone, after Gubin 1960) borders with the Tarim plateau on the east, with the Pamir Knot in the central part, and then stretches to the west to Gazli and further away. Focuses of the most violent earthquakes with M about 7.5 and more (Kashgar 1902, Karatag 1907, Khait 1941, Gazli 1976, 1984, Alay 1978, etc) are attributed to this zone.

The central Kyzyl-Kum fault zone has not been explored to a sufficient degree. Ibragimov et al. (1973) distinguish the flexular-fault zone of Kyzyl-Kum stretching in the sublatitudinal direction. Bezrodnyy et al. (1974) orient this zone to the south-west. It appears that both systems are correct and active in the seismic sense.

The western Tien Shan fault zone is not very much pronounced. It stretches in the south-west direction and separates the orogenic part from the Turan shield (Rezvoy 1962, Volvovskiy et al. 1966, and others).

The Pamir-Alay fault zone separates the Alpine folding of the Pamir Knot from Hercynids of southern Tien Shan; it stretches along Rivers Vakhsh, Surkhob and Kyzylsu at a distance about 400 km. It is extremely active in seismic terms.

The central Pamir fault zone consists of a series of sublatitudinal discontinuities. It separates the central Pamir from the southern and nothern Pamir.

The discontinuities described above, resulting from regional investigations, have referred to the major faults permitting categorization of the territory into highlands, mega-mesoblocks etc. The discontinuities of lower ranks are associated with separate geostructural units. A more detailed description of the latter is interesting in comprehensive investigations within geodynamical test areas. We will return to them in the chapters dealing with the tectonogeological characteristics of the test areas of Tashkent, Fergana and Kyzyl-Kum.

Profound structure of Middle Asia. The problem has been discussed by Arkhangelskiy & Fedynskiy (1936), Gamburtsev & Veytsman (1957), Ulomov (1966, 1974), Krestnikov & Nersesov (1962), Butovskaya et al. (1971), Savarenskiy et al. (1972), Kosminskaya (1960), Volvovskiy I.S. et al. (1962, 1966), Volvoskiy B.S. et al. (1973), Zunnunov (1973) and many others.

The investigation of the profound structure of the Pamir-Himalaya region has recently been conducted under an international programme in which scientists from the Soviet Union, India, Afghanistan and other countries have participated. The results of those investigations are presented in joint publications, for instance Anonymous (1982) etc.

From the data of numerous publications it can be concluded that the thickness of the Earth's crust in Middle Asia increases gradually from north-west to south-east. A direct correlation between the surface structural topography and the relief of the crustal foot is noted. It was already Arkhangelskiy and Fedynskiy (1936) who pointed to possible thickening of the Earth's crust in southern Tien Shan, as inferred from their analysis of the gravitational field.

The systems of the deep-reaching structure of Middle Asia proposed by various authors differ only by the configuration of isohypses; they all assume the increase in the Earth's crust thickness in the Pamir (Ulomov 1966, Borisov & Fedynskiy 1964, Melkanovitskiy 1965, Volvovskiy et al. 1956, Rezvoy 1964, Tal'-Virskiy 1964, Bulin 1972).

The orogenic area of Pamir-Tien Shan is characterized by gradual stepwise growth of the thickness of the Earth's crust from the northern borders (roughly from 40–45 km) to the south and south-east, in the direction of Pamir (to 60–70 km). Borisov (1964) singled out three steps: northern Tien Shan with depth variation to the Moho level from 40 to 50 km; Fergana from 45 to 50 km; and southern Tien Shan-Pamir, from 50 to 60 km and more.

A direct correspondence between the daylight topography and the granite layer level versus Moho has been observed for Fergana depression and a part of northern Tien Shan (Borisov 1967). The thickening of the crust occurs at a rate of 1 mm per year as postulated by Borisov (1967).

Results of the studies by Krestnikov & Nersesov (1962), Borisov (1967), Kosminskaya (1968), Butovskaya (1971), Ulomov (1966, 1974) and others have shown that the Earth's crust of Middle Asia has the block structure. The depth of the Moho level at the interface of the blocks changes in steps. The blocks are separated by profound faults. In the scheme of the deep structure of the eastern part of Middle Asia put forth by Ulomov (1974) one identifies interfaces of megablocks (northern Tien Shan, central Tien Shan, southern Tien Shan, Pamir & Kunlun, southern Pamir, Afghano-Tadzhik, and eastern Turan) and mesoblocks, and the location of basic profound faults. The megablocks are large and genetically homogeneous (Caledonian, Hercynian, Alpine, etc) structures of the Earth's crust having an area counted in hundred thousands of square kilometres. The profound faults on the peripheries of the megablocks stretch in hundreds of kilometres and penetrate the upper mantle in tens of kilometres, or sometimes cut across the entire lithosphere. The mesoblocks are components of megablocks. Their dimensions in plan vary from tens to hundreds of square kilometres, and by depth they encompass the entire thickness of the Earth's crust (Ulomov 1974, p.74).

3.2 NATURE OF GEOPHYSICAL FIELDS

The research on geophysical fields of Middle Asia was conducted by Arkhangelskiy, Belavskiy, Borisov, Gamburtsev, Mushketov, Nikiforov, Fedynskiy, Kosminskiy, Krestnikov, Nersesov, Volvovskiy, Ulomov, Tal'-Virskiy, Butovskaya, Fuzaylov, Yudakhin, Kurskeyev and others.

Gravity field. The anomalous field of the gravity force in Middle Asia is very specific. It reflects a variety of geological and tectonic factors such as the topography of the Earth's surface, profound structure, the composition and density parameters of rock components, the location of Paleozoic base, etc. It should be noted that the total field caused by different factors is depicted on charts of anomalous gravity field; thus identification of the field by a type of disturbance becomes an objective of special studies. In principle, depending on the formulation of the research task and a particular objective, the above problems are solved successfully in geological practice, so that we are not dwelling on them in this book.

The anomalous gravity field in Middle Asia in shown generally as a unigue gravitational depression with the centre within the Pamir Knot (Borisov 1967). The magnitude of the field decreases gradually from north-west to south-east, is weak to the west of the Turan shield and is characterized by an alternating band of positive and negative anomalies stretching in the direction NW-SE. Beginning from central Kyzyl-Kum the field is mostly negative. The gravity field in the area of Pamir-Tien Shan orogeny is characterized by intensive negative anomalies.

Taking into account that the depth of the Earth's crust in Middle Asia also increases from NW to SE, from 35–40 km in the west to 65–70 km in the orogenic part, and following Demenitskaya (1967, p.258) who believes that the major source of information on the thickness of the crust is contained in the topography of solid Earth and anomalous gravitational field (with proper account for regional relationships) one can claim that the general regional gravitational anomaly Δg is associated with the profound structure of the Earth's crust in the territory discussed. The smaller thickness of the crust corresponds to higher magnitude of the field, and vice versa, a thicker crust denotes a weaker field.

Anomalies of the consecutive rank include gravitational anomalies of regional and local character associated with the individual tectonic-structural units, the relief of basement and the composition of the latter, and also the Mohorovicic topography in some cases (Abdullabekov et al. 1980b). The anomalies have different forms: isometric, elongated along some particular structures, triangular etc. Their spatial dimensions also differ, their length ranging from 150 to 500 km, and the width measuring tens to hundreds of kilometres. The factors controlling the gravitational field are also diversified. Tal'-Virskiy (1972) has determined the following regularities:

a) Deep factors play a considerable role at the interface of the Turan shield and the orogenic area of the Tien Shan,

b) The petrographic inhomogeneity of the basement predominates in zones of the growth of magmatic formations,

c) The topography of the basement foundation has a prevailing effect in the regions where the basement consists of density-homogeneous sedimentary-metamorphic formations, and the surface itself is situated at a fairly small depth (up to 3 km),

d) The anomaly-generating boundaries move upward in the section due to the growth of density with depth towards the Pamir. Therefore in the region with deep basement the predominant effect on the formation of anomalies is caused by internal density interfaces, which are gradually higher stratigraphically, in the south-east direction.

The stretches of anomalies coincide mostly with the primary direction of the geological structures of the region. They are north-eastern in the northern Tien Shan, sublatitudinal in the central Tien Shan, and north-western in the Turan shield. The zones of higher gradients are attributed to fold disturbances. In intermontane depressions one encounters 'gravitational steps' associated with places of considerable gradients in the recent movements. As a rule, the places of high gradients of the gravitational field are located in seismogenic zones (Abdullabekov et al. 1980b).

In Borisov's opinion (1967), the zones of large gradients of the gravity field are associated with considerable changes (up to several kilometres) in the thickness of the entire Earth's crust, or to relationships between the 'granite' and 'basalt' layers.

Tal'-Virskiy (1964), Borisov (1967), Ulomov (1974) and others discriminate a zone of gravity field gradients between the Tien Shan orogenesis and the Turan shield. However, the gradiental zone distinguished on the chart of anomalous gravitational field is not distinct. Moreover, one cannot observe a dramatic thickening in the profound structure of the Earth's crust.

One also encounters gravitational anomalies of the third rank, i.e. local ones. They are associated with fine structural units, magmatic bodies, and other inhomogeneities of subsurface layers of the Earth's crust.

Properties of magnetic field. Some properties of the magnetic field of Middle Asia have been pinpointed by Borisov (1967):

a) The magnetic field of Pamir-Tien Shan, compared with the adjacent plains, is characterized by a weakly negative anomalous background. This corresponds to intensive processes of granitization and thickening of the Earth's crust during the period of Alpine tectogenesis,

b) The magnetic field of plain areas has some common features with the field of orogenic areas. In both territories one observes alternative positive and negative anomalies having a density counted in hundreds of nanoteslas. At the same time, the transition from the plain areas to the orogenic ones is gradual and smooth,

c) The orientation of the magnetic anomalies coincides with that of Paleozoic and Prepaleozoic folding structures. There is no relation to the orientation of the Alpine structure, which is attributed to the nature of the Alpine tectogenesis. In contrast to the Caledonian and Hercynian cycles of tectogenesis, the Alpine one was primarily accompanied by general transformation and disintegration of the crust without intrusion of magmatic substances of basic composition.

In the plain area of the territory one observes a well pronounced north-western and sublatitudinal folds, while a counterpart in the mountainous region is not clear and appears at random.

Studies by Tal'-Virskiy (1972), Fuzaylov (1977) and others contain information on regions of anomalous magnetic field and magnetic properties of rocks, together with a magnetic model of the Earth's crust for Middle Asia (and some other data).

The following magnetic anomalies are identified for Middle Asia (Fuzaylov 1977):

1. Areas of tranquil, large, and usually isometric (as to form) anomalies of different sign, of low intensity, with additional narrow local maxima and minima in some particular cases;

2. Areas with complex nature of the anomalous field, combinations of anomalies of different sign, configuration, orientation and intensity, which are grouped into zones of a certain direction in a number of cases;

3. Areas elongated linearly (belt-type), often with fairly intensive maxima, characterized by the clearly identified stretch. The latter often assume fairly small areas, and in such situations represent anomalous zones of the first two types. By sign, intensity, form, dimensions, stretch and mutual configuration of maxima and minima in Middle Asia one can single out a number of large anomalous areas and zones of magnetic field, which usually correspond to definite geostructural units.

The regional scheme put forth by Tal'-Virskiy (1972) includes extensive data on the form, dimensions, intensity and magnetic properties of large-scale magnetic anomalies belonging to individual magnetic bodies. The form and characteristic dimensions are most diversified. One can distinguish isometric, elongated, arched, mosaic and other forms having dimensions from hundreds to 700–800 km.

Tal'-Virskiy (1972) and Fuzaylov (1977) also confirm the presence of a direct or inverse relationship between the local anomalies of the magnetic and gravitational fields. The linearly elongated belts of maxima are identified with profound faults, which bring about disturbances in the Earth's crust reaching a considerable depth, and which generate favourable conditions for intrusion of magmatic melting.

The magnetic properties of rocks in Middle Asia have been studied by Abdul-layev, Akhmatov, Yeroshkin, Zarifbayev, Koriakin, Kunin, Lukashova, Maksu-dov, Melkanovitskiy, Orlovskiy, Rzhevskiy, Skovorodkin, Smelantsev, Tal'-Virskiy, Fuzaylov, Tsapenko and others.

The general findings on the magnetic properties of rocks (Fuzaylov 1977, p.128) can be summarized as follows:

1. Mesozoic-Cainozoic sedimentary rocks are weakly magnetic. The average magnetic susceptibility does not exceed 15×10^6 [cgs];

2. Sedimentary-metamorphic rocks of Paleozoic, and Riphean-Wend (lower subdivision of upper Proterozoic and uppermost Proterozoic, respectively) — sandstone, shales, limestones, etc — are also weakly magnetic. The mean value of their magnetic susceptibility is $20–30 \times 10^{-6}$ [cgs]. The exception is observed in the magnetic properties of the rocks situated in the zone of active contact with igneous rock (skarning, etc). Their magnetic susceptibility can reach tens of thousands $\times 10^{-6}$ [cgs];

3. The magnetic properties of Archaic-Middle Proterozoic metamorphic rocks vary in wide ranges from 0 to 3000×10^{-6} [cgs];

4. Magnetic properties of igneous rocks also vary in wide ranges, depending on a number of factors, among which the primary ones include the rock composition: the magnetism increases with increasing basicity.

Fuzaylov (1977) attempted a formulation of a geomagnetic model for the Earth's crust of Middle Asia. He established that the isothermal Curie surface had 30–35 km and was located below the 'basalt' layer.

3.3 CHARACTERISTICS OF UZBEK GEODYNAMICAL TEST AREAS

The Tashkent geodynamical area is situated in the zone of transition from the west Tien Shan orogenic area to the Turan shield. It is bordered on the east by Kuramin and Chatkal Ridges, while Karatau Ridge lies on its north, and the Nuratin and Malguzar Ridges are the southern boundaries. The Tashkent area has 10 000 km². The Paleozoic basement is overlain by sedimentary Mesozoic-Cenozoic deposis having a thickness of 1 to 3 km (Anonymous 1971). The basement rises gradually to the east, and becomes bare in the mountains of Karzhantau and at the south-west extremities of Chatkal Ridge.

The territory of the test area is characterized by an abundance of fault disturbances in different directions. Akhmedzhanov et al. (1971) group the latter by their spatial orientation into northwestern, sublatitudinal, northeastern, and submeridional. In the opinion of those authors, the primary network of faults is Proterozoic, or even earlier. Therefore they propose to classify them not by the time of generation but by epochs of intensive rejuvenation.

The faults of submeridional direction are not displayed in Paleozoic-Mesozoic; the northwestern directions are most intensive in Lower Paleozoic, the northeastern ones are seen in late Paleozoic and Alpine; they have been most active even recently.

Hence the 'youngest' faults are those in the north-east direction. The primary seismotectonic processes are attributed to this particular group of faults. The faults of this group are numerous in the Tashkent region but, as indicated by Ibragimov & Yakubov (1971), the highest seismic activity is displayed by faults of the Poltoratsk-Syr Darya seismotectonic zone being an extension of the Ugam-Karzhantau zone. It is the series of Alpine faults of the north-east orientation to which the epicentres of the strongest earthquakes of that region have been attributed (Pskem 1937, Brichmullin 1959, Tashkent 1966, Abaybazar 1971, Khalkabad 1972, Tavaksay 1977, Nazarbek 1980 and a number of weaker ones).

The northeastern system of faults is situated in both Mesozoic-Cenozoic and Paleozoic deposits. The Paleozoic basement of the test area, consiting primarily of intrusive, effusive and sedimentary-metamorphic formations is analogous by its lithological composition, to the Paleozoic outcrops of Chatkal-Kuramin mountains (Anonymous 1971).

Magnetic, electrical, elastic, radioactive, density and other properties of rocks have been investigated (Kuznetsov, Reshetov, Akhmatov, Melkanovitskiy, Boyko, Kunin, Tal'-Virskiy, Khvalovskiy, Kotlyarevskiy, Lapidus and others). All sedimentary rocks of the Mesozoic-Cenozoic age and the sedimentary-metamorphic

Table 3. Mean values of κ and I_n for constant intrusive rocks of Chatkal-Kuramin region, after Akhmatov (1970).

Rock	$\kappa x 10^{-6}$ [cgs]	$I_{nm} 10^{-6}$ [cgs]
Pyroxenites	10 600	1300
Gabbro	6050	1360
Syenito-diorites, diorites	2100–3000	200–430
Granodiorites	1400–1900	180–250
Granodiorite-porphyries	1100	260
Plagiogranites, adamellites	560–800	100
Leucocratic granites, alaskaites	23–170	8–20
Granites, porphyries, quartz porphyries	180–190	40–70

rocks of Paleozoic age are non-magnetic or weakly magnetic. The magnetic susceptibility and residual magnetism of these rocks are in the range of 10^{-6} [cgs].

Strong magnetic properties are displayed by igneous rocks of Paleozoic basement (Abdullabekov & Maksudov 1975). The data on the magnetic properties of magmatic rocks of Paleozoic age have been analysed so that the following conclusions may be drawn:

1. Acidic differences of effusions and their tuffs are weakly magnetic ($I_n = 100$–200×10^{-6} [cgs]). Tuffo-sandstones and tuffo-conglomerates are slightly more magnetic ($I_n = 60 \times 10^{-6}$ [cgs] on the average), while neutral and basic differences of effusions possess a fairly high magnetisation (100–1200×10^{-6} [cgs] on the average).

2. Among intrusive rocks, granites are weakly magnetic (with average $I_n = 200$–500×10^{-6} [cgs]). The remaining petrographic differences belong to magnetic and strongly magnetic rocks (mean $I_n = 2000$–3500×10^{-6} [cgs], Table 3).

The anomalous magnetic field of the Tashkent geodynamical test area is characterized by a complex combination of anomalies having different forms and intensities. The most distinct feature is their mosaic property. However, some elongated northwestern anomalies are fairly well pronounced inside this field of different signs.

Results of subaerial and ground measurements of magnetometric properties do not contradict the laboratory data. One encounters sharp alternating anomalies from -1000 to $+1000$ nT in the northeastern mountainous part of the test area, above mixed effusive formations; isometric forms of different density are usually observed above intrusions. Granites are characterized by slightly positive or negative anomalies, while granodiorites have primarily a positive field with a density of 400–700 nT (Table 4).

The basement of the Paleozoic plain part is submerged in a thick stratum of Mesozoic-Cenozoic sedimentary rocks. Therefore the magnitude of the anomalous field measures a few hundred nanoteslas.

Hence the Tashkent geodynamical test area possesses all properties permitting magnetometric studies such as

Table 4. Characteristics of magnetic anomalies above magmatic rocks of Paleozoic age in Chatkal subzone, after Kotlarevskiy (1969–1970).

Composition and age of rock	Magnitude of anomalies, nT			
	Karz	Chuli	Chir/Angr	SW Chat
Mixed effusions,	+300	+1000	-	− 150
P_{z3}	+600	+1000	-	− 150
Acid effusions	-	-	0	-
P_{z3}			0	−120
Granites, P_{z2-3}	0–150	-	−1000	+120
Granodiorites, P_{z2-3}	0–1000	-	+300	+400
			+1000	+500
Granodiorites	-	-	-	+200
P_{z1}				+400

(Karz) Karzhantau Ridge; (Chir/Angr) Basin between Rivers Chirchik and Angren; (SW Chat) SW of Chatkal Ridge.

a) The territory is fairly active from seismic point of view and includes specific seismogenic zones (Poltoratsk-Syr Darya and others);

b) Rocks of Paleozoic basement possess an optimum magnetism.

In order to confirm these conclusions we have computed the possible seismomagnetic effect for the Tashkent earthquake of 26th April 1966 (Abdullabekov & Golovkov 1970).

At the epicentre of earthquakes the Paleozoic basement is overlain by 2–3-km layer of Mesozoic-Cenozoic deposits. The basement is disintegrated by a network of faults — Hercynian in the north-west direction and Alpine of the north-east orientation. The Hercynian faults are filled with intrusions of basic and alkaline rocks with the magnetic susceptibility about 10^{-3} [cgs]. The elastic stresses are oriented along the blocks generated by the faults. In the belt between the faults, representing a narrow wedge, the longitudinal stresses could be relaxed only by oblique shearing and oblique-vertical dislocations. Such shearing obviously occurred during the Tashkent earthquake; this certainly took place at the interface of the filled postHercynian intrusion over the plain intersecting the edge towards north-west at an angle of $70°$ to horizontal.

In the case of Tashkent earthquakes the computations are much simpler because one can make a number of fairly well justified assumptions. The intrusion of strongly magnetic rock occurs in weakly magnetic rocks. The dimensions of the intrusion cutting across the wedge make it possible to assume that the entire structure is in the zone of homogeneous pressure, that is the compression from north-east to south-west. Assuming further that the directions of magnetic intrusion coincide with the direction of contemporary magnetic field (inclination of $60°$ and deviation below $5°$) one can take the projection of the magnetization vector parallel to the compression axis ($I_{\parallel}$) and the respective normal component ($I_{\perp}$):

$$I_p = 0.5I; \quad I_\| \approx 0.85I.$$

and

$$\Delta I_\| \approx 3 \times 10^{-5} \text{ [cgs]} \quad \text{and} \quad \Delta I_\perp \approx 40 \times 10^{-5} \text{ [cgs]}.$$

The total drop ΔI is 5×10^{-5} [cgs]. At the surface of intrusion this yields the magnetic field $H \approx 60$ nT.

At distances comparable with the depth of the upper edge of intrusion (≈ 2 km) and smaller depths of the lower edge (≈ 20 km), the reduction of the field in the first approximation reads $\frac{1}{R^2}$. If the characteristic dimension of the horizontal cross-section of the intrusion is 3 km, then the magnitude of the field at the distance of, for instance, 5 km will be about 20 nT.

A simpler evaluation can be provided on the basis of the following discussion. The entire magnetic body is in the zone of unique compression. The variation of magnetism inside this body is proportional to the magnetism itself. Since the magnitude of anomalies on daylight is also proportional to the magnetism, then one can roughly assume $\Delta T = T_a \delta C$. From aeromagnetic data one has the magnitude of the anomalous field about 300 nT, from which one has $\Delta T \approx 20$ nT, which can be detected by modern magnetometric apparatus.

One further investigated the variation of magnetic properties of the magmatic rocks of the Tashkent area subject to the variation of pressure and temperature, and detailed computations have been executed for the expected seismomagnetic effect for the Tashkent earthquake (Maksudov 1972, Maksudov et al. 1971, 1973). The computations with inclusions of the exposed relationships for specific rocks of the Tashkent area have shown that the magnitude of the seismomagnetic effect due to inductive magnetization was 0.3 nT, the real residual magnetization being 15.5, versus 388 for viscous and 17.5 nT for piezoresidual one (Maksudov 1972).

The Fergana geodynamical area is situated in the Fergana intermontane depression. The location is bordered by Kuramin and Chatkal in the north-west and north, Fergana in the east, and Alay and Turkestan Ridges in the south. The geological, geomorphological, tectonic, seismic, deep structural and geophysical features of the Fergana depression have been described in many studies (Ibragimov 1970, 1978, Anonymous 1982, Yarmukhamedov et al. 1979, Yudakhin 1983, Volvovskiy et al. 1962, Khamrabayev 1977, and others).

The intermontane depression of Fergana and its mountainous borders consist of Proterozoic, Paleozoic and Mesozoic-Cenozoic rocks. The rock genesis involves sedimentary, magmatic and metamorphic complexes.

Precambrian and Paleozoic deposits in the form of metamorphic, sedimentary, volcanogenic and intrusive rocks of neutral and acidic composition are found at the surface of the rock border. Silurian, Devonian and Carboniferous rocks in limestones, sandstones, shales, conglomerates and volcanogenic formations are widespread in the depression. Permian rocks are rarer, in carbonate-terrigenic deposits and interlayers of acid effusions.

The thickness of Mesozoic-Cenozoic deposits increases from the mountainous part to the centre of the depression and reaches 8–9 km or 12 km at places. They consist of limestones, dolomites, clay, sandstones, gravelites, aleurolites, conglomerates, loam clay and other sedimentary rocks.

The Earth's crust of the considered region changed considerably in the recent geological times. The general drop of altitudes of the Earth's surface between the zones of rises and descents is 10–12 km. For example, Ibragimov (1972) indicates that the amplitudes of the recent ascending movements in Kuramin and Chatkal Ridges has been 4 km, versus 4–5 km for Fergana Ridge, 6.5 km for Alay and Turkestan Ridges, and 3 km for the amplitude of descent in the Chatkal depression versus 7 km in the Fergana depression.

The Fergana area is also characterized by an abundance of discontinuities and a high seismic activity. The most active areas are southern Fergana, Kurshab, northern Fergana, Chatkal-Atoynak, Talas-Fergana faults and southern Fergana and northern Fergana flexural-fault zones (Ibragimov 1978). The focuses of the most violent earthquakes (Omsk 1883, M=7.75; Leninabad 1886, M=6; Kostakoz 1888, M=5.5; Chust 1894, M=5.75; Ura-Tyubin 1887, M=6.5; Andizhan 1902, M=7.5; Alim 1903, M=6.5; Kyrkkol' 1907, M=6.5; Kurshab 1924, M=6.5; Dzhalalabad 1926, M=6.75; Namangan 1927, M=5.75; Karakalpak 1947, M=5.5; Markay 1962, M=5.5; Supetau 1967, M=5.25; Isfara-Batken 1977, M=5.4; Khaydarken 1977, M=5.7; and others) are attributed to those discontinuities (Ibragimov & Abdullabekov 1974a).

As a rule, a combination of stretches of distinct recent and contemporary movements of the Earth's crust, zones of profound faults and high gradients of geophysical fields, and focuses of violent earthquakes generate seismogenic zones. Ibragimov (1978) discriminated three categories of seismogenic zones in the Fergana depression for 9, 8, and 7-degree quakes. The first category incorporates northern Fergana, Namangan and Andizhan seismogenic zones. They stretch as narrow belts from north-east to south-west, and merge in the west with zones of the second category. The latter include Uratyub-Kurshab, Samgor, Kostakoz, Kukumbay-Abadan and other zones. The seismogenic zones of the third category (Kokand, Izbaskan and others) are situated separately to the east and west of the depression.

The magnetic parameters of the rocks of the depression vary in wide ranges shown in Table 5. The tabulated data shows that the Mesozoic-Cenozoic deposits of the depression are weakly magnetic or non-magnetic (10–50 $\times 10^{-6}$ [cgs]). The Premesozoic formations are also weakly magnetic. The magnetic properties of the igneous rock of Paleozoic and Prepaleozoic vary from 310 to 10000 $\times 10^{-6}$ [cgs].

Anomalous magnetic field of Fergana depression.   Subaeral and aeromagnetic surveying in the Fergana Valley has been conducted since 1932 (Bezkrovnyy, Zilbershtein, Kotlarevskiy, Kremnev, Kuznetsov, Lomanov, Mingel, Reshetov and others). Until now a fine scale chart of isolines for ΔT has been complied for

Table 5. Summary of data on magnetic susceptibility of Fergana depression rock after Lukasheva & Kharitonov (1963).

Age	W Ferg	S Ferg	SE Ferg	E Ferg	NE Ferg	Aver.
Neogene	$\frac{25.8}{374}$	$\frac{32.6}{13}$	$\frac{62.6}{59}$	$\frac{28.3}{6}$	$\frac{27.2}{1184}$	$\frac{34.9}{636}$
Paleogene	$\frac{10.7}{532}$	$\frac{17.5}{469}$	$\frac{33.6}{112}$	$\frac{28.1}{98}$	$\frac{15.9}{407}$	$\frac{21.5}{1613}$
Cretaceous	$\frac{7.8}{369}$	$\frac{12.2}{159}$	$\frac{27.2}{305}$	$\frac{25.7}{396}$	$\frac{16.3}{173}$	$\frac{20.2}{1407}$
Jurassic	$\frac{35.9}{1200}$	$\frac{25.3}{1}$	$\frac{77.7}{9}$	-	$\frac{76.5}{72}$	$\frac{51.1}{1282}$
Paleozoic	$\frac{3.1}{327}$	-	-	-	$\frac{62.7}{27}$	$\frac{53.2}{354}$

The numerator denotes the magnetic susceptibility multiplied by 10^{-6} [cgs], while the denominator indicates the amount of samples.

the territory of Middle Asia (Chart of magnetic anomalies, 1970), aside from a medium-scale chart of ΔT isolines for Fergana depression and a large-scale chart for individual segments of the valley (Fig.17). The given chart of the test area is characterized by a somehow higher magnetic field. The numerous local anomalies are confined in the depression stretching from the south, east and north-east; it is only in the north-west direction that intensive positive anomalies of the Chatkal-Kuramin Ridge appear.

The largest anomalies in the depression are identified as Kokand, Maylisuy and Markhamat, with a density reaching a few hundred nanoteslas; they are accompanied by finer positive and negative magnetic anomalies.

Kyzyl-Kum test area. In seismotectonic terms, the territory of the central Kyzyl-Kum has been explored in fewer details (Ibragimov et al. 1973). A disperse network of seismic stations has not yet permitted an elaboration of extensive data for seismic characteristics of the territory. However, the intensive phenomenon of fissuring which began in the sixties, together with the violent Gazli earthquake on 8th April (M = 7) and 17th May (M = 7.3) 1976, with the later earthquake on 20th March 1984 (M = 7.1) have forced the researchers to evaluate the seismotectonic characteristics of the region.

The chart of seismic regions (1983) places the area of the central Kyzyl-Kum in the category of 7–8-degree earthquake zone. It has been proved that the central Kyzyl-Kum is not inferior to the orogenic part with respect to the seismicity and the tectonic activity. The celerities of contemporary movements of the Earth's crust are also close to each other in the two regions.

The Kyzyl-Kum area is a favourable site for magnetometric studies in view of the absence of industrial and other noises and the low field gradient, which largely simplify the field measurements. The magnitude of the anomalous field along the deployed routes does not exceed 100–150 nT (Fig.18). Any substantial local and regional anomalies associated with intrusive bodies and structural properties of the region are not manifested.

Figure 17. Chart of ΔT_a isolines for Fergana depression after Kotlarevskiy & Kremnev (1975): (1) positive; (2) neutral; (3) positive; in nT.

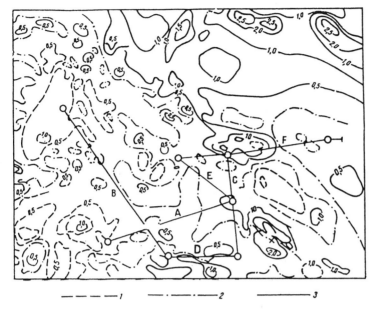

Figure 18. Chart of ΔT_a isolines (in millioersteds) for central Kyzyl-Kum, copied from Chart of magnetic anomalies of the central part of Middle Asia (edited by Tal'-Virskiy & Fuzaylov 1970). (A-F) routes of repeated magnetic measurements; (1) positive; (2) neutral; (3) negative.

The magnetic properties of the rock in the region discussed have been little explored. Judging from bibliograpbhical sources, the highest magnetism is exhibited by the rocks of Paleozoic basement (5×10^{-3} [cgs]). At the outcrops of Paleozoic rocks they display magnetic anomalies up to 40–50 nT. However, along the primary routes, the Paleozoic formations are overlain by Mesozoic-Cenozoic formations, so there is no intensive anomaly of the geomagnetic field.

CHAPTER 4

Techniques of magnetometric investigations in Uzbekistan

4.1 SURVEY PARAMETERS

Anomalous changes in the geomagnetic field associated with processes in the Earth's crust, i.e. magnetic effects, can be characterized by three parameters:

$$\Delta T = f(t, L, T) \tag{4.1}$$

in which

$\quad\quad t =$ duration of effect,
$\quad\quad L =$ linear dimension,
$\quad\quad T =$ field density.

Measurements of the magnetic field in test areas consist in determination of the above three parameters by a configuration of measuring points in space, which can generally be described by a certain parameter l, being an average spacing of measuring points, together with the frequency of repetition Δt and accuracy of measurements δT, which is usually given in terms of the standard deviation σ.

In detailed investigations of anomalous changes in the geomagnetic field, the characteristics of the measuring network should be smaller at least by one order of magnitude than those of the measured effect. Hence the selection of a measuring technique is reduced to the evaluation of t, L and T and determination of respective characteristics of the network $\Delta t, l$ and δT.

Basing on an analysis of the magnitude of the excess elastic stresses in the Earth's crust, the spatial and temporal dimensions of tectonic movements, seismicity, literature findings, and results of laboratory modelling and field investigations on tectogenic objects (see Chapters 1 and 2), one can provide an approximate assessment of the expected magnitude of the parameters pertaining to magnetic effects.

The duration of the effect varies from minutes to tens of years. At the same time, the temporal characteristics can be classified in the three groups: slow, transitional (mesoscale) and fast.

Slow changes (tens of years) are primarily associated with contemporary movements of the Earth's crust and some other processes. One can place in this category the properties of the secular change in the magnetic field at observatories, the anomalous changes discovered in the Ural test area and other factors (Golovkov et al. 1977, Abdullabekov & Golovkov 1974, Shapiro 1982, 1983, and others).

The transitional (mesoscale) changes of the geomagnetic field are associated with the processes of elastic stress build-up during earthquakes, incipient eruption of volcanos, subsequent relaxation of elastic stresses and temperatures, etc. The duration of this type of variation ranges from one month to some years, the characteristic dimensions being counted in tens of kilometres, and the density varying from fractions and units to the first tens of nanoteslas. The characteristic times of the deformation changes in the Earth's crust, inclinations of the Earth's surface, β-movements of the Earth's crust, sea level and the higher seismic activity prior to volcanic eruptions are also in the above range (Golovkov et al. 1977, Abdullabekov 1972, Abdullabekov & Maksudov 1975, and others).

Fast changes in the geomagnetic field, with characteristic times from days to 2–3 weeks are obviously associated with calamitous processes in the Earth's crust due to the generation of major fissures and the liberation of energy during earthquakes, eruptions of volcanos, movements of masses, etc. The existence of such times in the physics of the Earth is also testified by effects confirmed by other methods such as S-type anomalous changes in the inclination of the Earth's surface prior to earthquakes, γ-movements of the Earth's crust due to earthquakes (Meshcheryakov 1968b), anomalous contents of radon and gaso-chemical composition in thermo-mineral waters before earthquakes, and many other effects having the same characteristic times, so typical for fast changes in the geomagnetic field (Zubkov & Migunov 1975, Golovkov et al. 1977, Rikitake 1979, and others).

Analysis of the results of field measurements and theoretical studies, together with observatory data (see Chapters 1 and 2) shows that the spatial dimensions of the field changes are diversified due to their different nature. One can prove that the characteristic dimensions of transitional and fast variations are similar, i.e. tens or first hundreds of kilometres. The linear dimension of the slow field variation reaches hundreds of kilometres, as can be inferred from observatory data and the results of area surveying in the Ural test area (Shapiro 1982, 1983), and in Japan (Tazima 1968).

The density of anomalies for effects having different characteristic times is also diversified: fractions and units of nanotesla for fast changes, units and tens for transitional ones and 50–70 nT for slow changes.

With inclusion of the above information on the parameters of the effects discussed and using bibliographical data (Ulomov 1977, Dobrovolskiy 1980) we have assessed parameters of the network of repeated area, route and stationary measurements (Table 6).

The technique of continuous measurements or repeated surveying is clearly identified in the table. Fast changes can be quantified by continuous measurements. The first measurements of this type were carried out by Breiner & Kovach (1968). Table 6 is noteworthy as it permits selection of surveying parameters, which depend

Table 6. Approximate dimensions of magnetic effect parameters and respective characteristics of measuring network.

Magnetic effects	Time, days		Dimensions, km		Density, nT	
	t	Δt	L	l	T	ΔT
Slow	$> 10^4$	$\approx 10^3$	$n \times 10$	10	below $n \times 10^2$	10
Transitional	30–10^3	≈ 10–100	$n \cdot 10$	1–5	below $n \times 10$	1
Fast	1–20	0.1–1	$n \cdot 10$	$< n \times 1$	< 10	0.1

primarily on an objective. If the latter is a search for transitional or slow field changes, then it is sufficient if at least one point of the network coincides with the location of the anomalous field changes.

In detailed investigations it is necessary that the zone of the effect be divided by a number of sections. The increment of surveying should be smaller by one order of magnitude than the linear dimension of the effect.

With reference to the frequency of measurements, the condition $\Delta t \leq 0.1\ t$ given in Table 6 is fully compatible with the temporal field change of a monotonic process. In practice, for the transitional processes, the frequency of measurement should range from several days to a few months, while years are selected for slow processes.

In assessment of the zone of occurrence of different earthquake forerunners (precursory effects) it is most suitable to take the relation $R = e^M$ km (Dobrovolskiy et al. 1980). Table 7 (Ulomov 1977) provides dimensions of the earthquake focus (P_1) and respective ten (P_{10}), twenty (P_{20}) and thirty (P_{30}) focus dimensions. The experience of field studies shows that anomalous forerunners occur at distances up to 30 focus dimensions. In some cases the effects are also observed at larger distances.

Starting from the relationship $R = e^M$ km, an optimum network of measuring points can be selected from Table 7. The increment of repeated cycles of measurements is determined from the known relationships between the time of appearance of different effects versus earthquake magnitude. The graphical representation is given for the logarithms of time (days) versus the magnitude and energy class of earthquake (Myachkin et al. 1975, Sidorin 1979, Whitcomb et al. 1973, Rikitaki 1979, Sholz et al. 1973, and others). From this data it follows that the duration of anomalous changes for earthquakes with, for instance, M = 5 ranges from 56 to 176 days, while the characteristic time for the effects due to stronger earthquakes is even longer.

Basing on the expected magnitude of anomalous changes in the magnetic field one selects the surveying parameters: spacing of measuring points, frequency of surveying and accuracy of measurements. Tie field works have aimed at identification of the transitional and fast field changes. The transitional (mesoscale) changes were investigated in networks of repeated route surveying in the Tashkent, Fergana

Table 7. Relationships between energy class (K), magnitude (M) and size of earthquake focus (P, km), after Ulomov (1977).

K	M	P_1	P_{10}	P_{20}	P_{30}
9	2.8	0.3	3	6	8
10	3.3	0.5	6	11	16
11	3.9	1.1	11	22	33
12	4.4	2.2	22	44	66
13	5.0	4.4	44	88	131
14	5.6	8.8	88	175	262
15	6.1	17.5	175	360	525
16	6.7	35	350	700	1050
17	7.2	70	700	1400	2100
18	7.8	140	1400	2800	4200

and Kyzyl-Kum test areas, three to four times a year, in addition to the system of forecast stations. The fast changes were monitored by routine measurements in a network of stationary stations.

4.2 TECHNIQUE OF MAGNETOMETRIC OBSERVATIONS IN TEST AREAS: CONFIGURATION OF MEASURING POINTS

The routes of repeated surveying in the Uzbek test areas have been arranged along primary seismogenic zones, deep-reaching faults and flexural-fault zones, with a surveying increment of 4–5 km.

In selection of the route one took into account the geological and tectonic situation, the layout of profound active faults and seismogenic zones, and areas of intensive contemporary movements of the Earth's crust, etc.

In investigation of the seismomagnetic effect the success of the study depends largely on the selection of objects. Considering the seismotectonic conditions of the region we selected as the primary objects of investigations the Tashkent and Kuzyl-Kum geodynamical areas, along with Charvak Reservoir and the artificial underground gas storage mentioned above.

In the Tashkent geodynamical area we extended two routes along the Karzhan-tau Fault, a third one across them, and three additional routes in the regions of Charvak Reservoir, Poltoratskiy gas storage and between the populated sites of Sary-Agach-Chinaz. The routes in the Fergana test area passed along the northern and southern Fergana faults; they embodied numerous seismogenic zones. The routes in the territory of the central Kyzyl-Kum area have been arranged in sections of intensive contemporary movements of the Earth's crust and in the epicentral zone of the Gazli earthquake of 1976.

The points of stationary routine measurements have been located in territories of extensive forecast stations or in their close proximity. The forecast stations were also situated in zones of active seismotectonic conditions.

Error analysis. In order to eliminate the variations of magnetospheric and ionospheric origin, all measurements (both for points of repeated measurements and at stationary stations) have been referred to a common point, so that the instantaneous differences of field values were always taken for two points (reference and a given route point).

The standard deviation of measurements at a point (σ) for each region varies as a function of the geographical location, geological structure, applicability of measuring techniques, apparatus and a number of other factors. It is determined by the instrumental error of the proton magnetometer (σ_{pr}) and incompatibility of the variation at two sufficiently remote points (σ_{var}) and observatory (σ_{obs}):

$$\sigma = \sqrt{\sigma_{pr}^2 + \sigma_{var}^2 - \sigma_{obs}^2}. \tag{4.2}$$

In the technology od synchronous measurements at reference and a given point or at two stationary points the quantity σ is determined by the error of two proton magnetometers and the difference of the variation course at two points:

$$\sigma = \sqrt{2\sigma_{pr}^2 + \sigma_{var}^2}. \tag{4.3}$$

Instrumentation error. The certified accuracy of measurement by PM-5 proton magnetometers is 1.5–2.5 nT, versus 0.1 to 0.2 nT for TMP-1, MPP-1 and MPP-IM and 1–1.5 nT for PMP, PM-001, M-32 and Zh-816. The above magnetometers are subject to random and systematic errors, of which the former are eliminated automatically in a series of records while the latter are divided into permanent and temporary. The permanent systematic errors are linked to inaccurate determination of the frequency by quartz oscillators, gyromagnetic relation and some other properties of instrumental circuitry. The variable systematic errors are caused by conditions of operation, that is temperature, voltage of sources, ageing of circuitry, signal level, etc.

The permanent instrument errors can be included upon combination of several instruments of one type or by reference to a standard instrument. It is only through detailed examination of each instrument that its individual standard deviation (σ_{ins}) can be found. To this end one intercalibrates a few proton magnetometers by comparing a series of not less than 150–200 records per each instrument. Subsequently the tests on the instruments are carried out in the ranges of working temperatures which will be encountered in field studies.

In order to check on the identity of indications by magnetometric apparatus of different types, discrete measurements every 5–10 minutes were conducted at Yangibazar Observatory in a course of a few days (Fig.19). The drawing shows that the readings of all magnetometers were identical.

Similar studies were conducted for each measuring instrument several times a year so that instrument errors were determined from each magnetometer, ide-

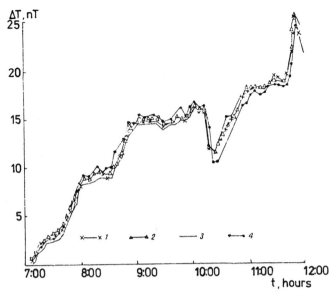

Figure 19. Graphs of simultaneous readings of different types of magnetometers at Yangibazar Observatory: (1) Helium-1; (2) Helium-2; (3) M-32; (4) MVS.

pendently of the certified data. The real accuracy of the applied apparatus was evaluated as

$$\sigma_{ins} = \sqrt{\frac{\sum \Delta T^2}{n}} \qquad (4.4)$$

in which

    $\Delta T=$ deviation from the mean value of several instruments,

    $n =$ number of differences.

The quantity σ_{ins} was determined by numerous tests for different types of magnetometers used in our test areas. The standard deviation of the instrument error was as follows: 1.5–2 nT for PM-5; 1–1.5 nT for PMP, PM-01, M-32, Zh-816 and MPP-203, and 0.2–0.3 nT for TNP, MPP-1 and MPP-1M.

    The error of field measurement by records of a magnetovariational station (MVS) consists of the reproduction error of variation and the measurement error of the station reference. The deviation of the reading at MVS from the real course of variations is determined by errors due to orientation, sensitivity, effect of moving parts of recording apparatus, record jumps upon reloading of the recording instruments, time service errors and some other causes. If all requirements imposed by MVS instructions are observed, the error σ_{MVS} does not exceed 1.5–2 nT, although some stations can suffer from technological defects. The real standard deviation σ_{MVS} is given as follows:

$$\sigma_{MVS} = \sqrt{\frac{\sum \Delta T^2}{n}} \qquad (4.5)$$

in which

Δ = difference of instrument readings,

n = number of data sets.

The errors of reference MVS values are associated with incorrect arrangement of the station and a considerable daily variation of temperature in the measuring room. Higher errors can be avoided if the following relationship is adhered to:

$$\Delta t^\circ\, E < 1\ \text{nT} \qquad (4.6)$$

in which

Δt° = daily variation of temperature in measuring room,

E = temperature coefficient for the station.

In determination of the possible errors of the variation service at Yangibazar Observatory we have used the data of simultaneous measurements of T with the aid of helium and proton magnetometers. The measurement error ΔT computed by comparison of a T-variometer with helium and proton magnetometers was 0.4–0.5 nT. The quantity σ_{MVS} also remained constant in the next years.

Upon subtracting the data for the reference station and the given point we always considered the time of subtraction; if it occurred at an instant of a dramatic field change the results were corrected or rejected.

Errors caused by the variation difference (σ_{var}). These errors can be caused by three factors: (a) if the point is at a different distance from the source of variation, so that the field variation at both stations is different; (b) the geomagnetic variations induce electric currents which distort the variations at a given point if there appear local peculiarities of electrical conductivity of the Earth's crust and upper mantle; (c) upon investigation at industrial objects one can encounter industrial noises.

Let us evaluate the difference of variations due to the first cause. This can be done by the example of S_q variations. Their amplitude in element T of the Soviet territory can reach 50 nT, and its rate of growth is up to 0.3 nT/min. Since the variations depend on local time, they do not appear at the same time at two points spaced latitudinally (the occurrence will be earlier at the eastern point). For the southern regions of the Soviet Union (ϕ =40–45°) the quantity σ_{var} at a spacing of 100 km is about 1 nT.

Hence if the error of S_q variation has to be lower than the instrument error, the greatest spacing between the reference and the measuring point in the territories of the geodynamical test areas of Uzbekistan should be below 100 km. Taking this into account, the measurements were related to Yangibazar Observatory in the Tashkent test area only, while the method of synchronous measurements at a reference and measuring points was employed in the Kyzyl-Kum and Fergana test areas. The characteristic courses of variation at Yangibazar Observatory and the reference station of Tamdybulak are identical. Similar results have also been obtained at the remaining sections of the Kyzyl-Kum test area. The differences observed seem to be caused by the dissimilar geographical configuration of the observation points. It is known that the daily solar S_q variations of the field at easterly points occur earlier. At the time of maximum variation the gradient of field change at the latitude of Yangibazar and Tamdybulak reaches 0.5 nT/min. The distance from Yangibazar to Tamdybulak is 450 km, which makes a difference of 15 minutes. Hence the field difference due to the above variation can amount to 7–8 nT.

Local properties of the course of variations caused by inductive effects in conducting rock volumes are well known; they are used in electrical prospecting by the method of magnetovariational probing (Yanovskiy 1963, Rikitaki 1968, and others). Let us consider the technique of inclusion and elimination of the errors which could be caused by the induction effect in the measurements as related to the search for the seismomagnetic effect.

During extremely strong disturbances in the geomagnetic field (magnetic storms), the difference of data at two points spaced by 70–100 km can reach tens of nanoteslas, that is by two orders of magnitude above the instrument error. Therefore the results of measurements during strong disturbances have been rejected and excluded from further processing. Primary attention was paid to identification of the places where the induction distortion of the variations reached a considerable magnitude.

Studies on the course of S_q-variations were conducted in geodynamical test areas of Uzbekistan. The first observations on S_q in the Tashkent area were initiated in the years 1969–1970 (Abdullabekov 1972). Daily variations of the geomagnetic field were measured over a few days at different segments of the test area. It was found that zones with anomalous electrical conductivity, which would distort the course of S_q-variations did not exist within the Tashkent area. The quantity σ_{var} for the most remote points of the area (south-western, western, northern and north-eastern part of the area) did not exceed 1 nT.

In recent years, due to availability of high-sensitivity magnetometers and a higher accuracy of magnetic surveying, detailed measurements of S_q-variations were conducted in the Tashkent area. Individual segments with σ_{var} up to 0.5–0.6 nT due to distortions of inductive nature have been identified within the south-western part of the area (Tsvetkov et al. 1982).

The solar-day variation of magnetic field was also investigated in the Kyzyl-Kum area. Measurements of the field variations were conducted at a number of points. The quantity σ_{var} between the reference of Tamdybulak and the most remote points of the area during calm days was 0.5–0.7 nT, versus 0.7–0.9 for days with an interference.

The most extensive studies on the behaviour of S_q-variations were conducted in the Fergana geodynamical area. A number of reference and regular points (point 103, references of Yaz"yavan, Shakhimardan, etc) were observed under special programmes on the variation of the geomagnetic field, and the results were compared against the data measured at Yangibazar Observatory.

At point MP-103 (Sovetabad) the difference at the distance of 120 km reached up to 5–8 nT. The amplitude distortions of bay-type disturbances were up to 5–8 nT. The amplitude of the variation at MP-103, despite its location to the south of Yangibazar Observatory (120 km) was almost two times higher than the amplitude of the variation at the observatory. These large differences in the variation are attributed to inhomogeneity of the electrical conductivity of the rock in the territory of measurements.

The reference point of Yaz"yavan is located in the central part of the Fergana depression. The region is characterized by a calm magnetic field (ΔT_a). The S_q-variations were measured from 5th to 7th August 1975. They have shown amplitudes of individual bays with a period of 1–3 hours being up to 20–25 nT. The deviations of differences from the mean value reached up to 8–10 nT at extremum segments of the curve. Therefore one can conclude that the Yaz"yavan region is also characterized by an anomalous electrical conductivity.

Investigation of the behaviour of variations about the station of Shakhimardan was conducted during a relatively calm time of the year (3rd-4th December 1976). At that time the amplitude of S_q-variations was the lowest, from a few to the first tens of nanoteslas. The variations recorded on the 4th December 1976 can be taken as an example. On that day the magnitude of the field increased gradually and turned into disturbed conditions, exhibited as sinusoidal oscillations with gradual alternation. The amplitude of the variations between 2:00 p.m. and 3:20 p.m. at Yangibazar was 22 nT, versus 33.0 nT at Shakhimardan.

Judging from results of the investigations of the variations, the Shakhimardan station was also located in a zone of anomalous electrical conductivity. In such zones, fast changes in the direction and magnitude of the magnetic field bring about induction currents, which in turn produce an additional field coinciding as to direction with the external field. The magnitude of the standard deviation of the difference during a disturbed day (4th December) was ±1.9 nT.

The measurement of the magnetic field variations at Batken was conducted after the Isfara-Batken earthquake on the 31st January and 24th-26th August 1977 (Fig.20). The drawing clearly shows higher amplitudes of the variations at Batken, compared with Yangibazar. During the time of bay passage (from 5:10 p.m. to 7:30 p.m.) the deviation of the differences from the mean value reached up to 13.3 nT, that is the amplitude of the bay at Batken was two times higher than that at Yangibazar.

The magnitude of the standard deviation for the difference in the case of a disturbed day (25th March 1977) was ±5.3 nT. In order to identify and analyse the variations of the geomagnetic field associated with earthquakes, it was necessary to carry out magnetic surveying within an accuracy higher than 1 nT, as stipulated by recommendations (Golovkov et al. 1977).

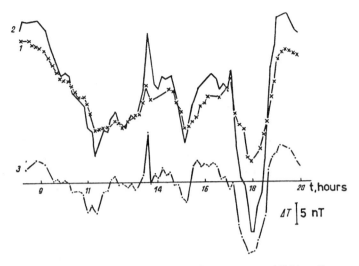

Figure 20. Course of S_q-variations at Batken in Fergana area: (1) Yangibazar; (2) Batken; (3) ΔT Batken-Yangibazar.

Hence one can conclude that the type of S_q-variations in the Fergana geodynamical test area is more complex than that within the Tashkent and Kyzyl-Kum areas. Particularly high distortions were observed in southern Fergana along the zone of southern Fergana deep-reaching fault and the same seismogenic zone, where induction currents linked to the anomalous electrical conductivity rock bring about a considerable contribution to S_q-variations. The same effect, although less pronounced, was exposed within the central Fergana (Yaz"yavan). In a number of cases, the substantial variations of induction type have been attributed to seismogenic zones. Therefore one must pay particular attention to the variations of the induction origin, if the seismomagnetic effect is investigated.

Hence we have considered all types of possible errors encountered in investigations of the seismomagnetic effect in test areas.

Basing on the methods of observations applied we have determined standard deviations for the measurement errors within each test area. In the Tashkent area the quantity σ for measurements at regular points with reference to the base line of Yangibazar T-variometer was 1.5–2.0 nT. Since 1975 the measurements have been conducted by the method of synchronous observations, in which the standard deviation of the error was about 1 nT.

The quantity σ in the Fergana area was up to ± 3 nT until 1975. Because a new method of measurements and a new type of apparatus (TM-P with a sensitivity of 0.1–0.2 nT) have been used since 1975, the accuracy of surveying increased to 1–1.2 nT.

Even a smaller value has been obtained for the east Fergana area, where repeated measurements were employed to evaluate the real accuracy of measurements; this was done during 2–3 days at different times of day. Results of double idependent measurements in series at those points have produced the standard deviation of the error of 0.6–0.7 nT. The counterpart of the error for the Kyzyl-Kum area was 1.4–1.8 nT.

Assessment of accuracy of stationary measurements of the magnetic field. If the seismomagnetic effect is measured with the MPP-1, MPP-1M or APM station- ary absolute proton variometers, then one can identify the effects with different characteristic times. If these phenomena are short-term, so that their duration is only several times longer than the time step between discrete measurements at the stations, the problem of the error is solved as for the discrete measurements at individual points of routes. If however one is interested in a phenomenon lasting a few days or more, the use of stationary stations makes it possible to reduce sub- stantially the error because the quantity of regular measurements increases (the statistics does so) and due to different courses of the variations σ_{var}.

As shown, aside from magnetic storms, the highest contribution to σ_{var} is caused by the bays. Accordingly, even for simple averaging (aimed at computation of mean hourly or mean daily values) the contribution of the error decreases proportionally to the ratio of the duration of the bay to the time of averaging.

The accuracy of measurements for each station and for couples of close stations is evaluted by the formula:

$$\sigma = \sqrt{\frac{\sum \delta T^2}{n}} \qquad (4.7)$$

in which

σ_{pr} = accuracy of stationary magnetometer,

σ_{var} = difference of variations between two stations.

As can be seen, the formula for the accuracy of stationary measurements is identical with the formula for regular synchronous measurements in test areas.

The certified accuracy of a single measurement of the magnetic field using the magnetometer MPP-1, MPP-1M and APM is 0.1–0.3 nT. In order to find the real accuracy of the magnetometers applied one must conduct regular assessment studies. To this end, the same place is used at which a few stations are arranged, and the synchronous measurements are conducted over 2–3 days. The standard deviation for each magnetometer is given as follows:

$$\sigma_{pr} = \sqrt{\frac{\sum \delta T^2}{n}} \qquad (4.8)$$

in which

σ_{pr} = standard deviation of accuracy for proton magnetometer,
δT = deviation of reading of a tested magnetometer related to mean value,
n = number of synchronous measurements.

Similar studies were conducted at each station a few times per year. As a rule, the accuracy of the magnetometers employed was not higher than 0.2–0.25 nT.

The accuracy of the measurement of the field difference ΔT was assessed for each station and Yangibazar Observatory, and also for couples of adjacent stations. The standard deviation of the accuracy for mean daily estimates, referenced to the data of Yangibazar Observatory for the Tashkent area was 0.3–0.6 nT, while its counterpart for the Fergana and Kyzyl-Kum areas was 1–1.5 nT, versus 0.4–0.7 nT for couples of close stations in the eastern Fergana area. Upon construction of time series for the differences we have encountered some specific difficulties:

a) For identification of the effect it was convenient to use differences for remote stations as the effect decreases with increasing distance. In construction of differences for field variations we identified not the absolute effect but a certain gradient between the stations — the greater the spacing between the stations the higher is the effect;

b) It is known that the standard deviation also increases with distance. For instance, σ for couples of close stations in the Andizhan area was 0.4–0.8 nT while that between the stations of the Andizhan area and Yangibazar Observatory was above 1 nT. Taking into account these extremities it is aapropriate to utilize optimum distances between pairs of stations.

In studies on anomalous effects in the geomagnetic field due to processes in the Earth's crust the magnitude of the effect is often within the range of the first units of nanotesla. Therefore it is necessary to attempt to increase the accuracy of the identified anomalies. A few tools are available:

1. Higher quantity of measurements;
2. Use of time filters;
3. Use of convenient time of day during measurements.

One also knows some other possibilities of reducing the error of the mean daily differences which are based on the fact that their cause is associated with the disparity of the time of variations at two stations. Consider a series of mean daily magnitudes of the field and the corresponding series of differences. If the series are correlated, the series of differences contain errors due to the different course of the variations. This can be excluded if one determines the regression coefficient and subtracts the correlated part from the series of differences. This procedure is applied if the series contain a fairly high amount of data (not less than 20), i.e. the method can be employed for identification of long-term effects.

For shorter series we have proposed and tested a method of modal daily values, which consists in the procedure within which for every day one constructs a histogram of mean hourly (or average values for lapses of 20 minutes) differences for station couples, and the modal value of the resulting distribution is taken as the real difference. The method is close by its concept to the estimation of the mean monthly value by five calm days (or by night time) in the practice of observatory measurements.

The two methods described permit reduction of the measurement error for the differences to 0.4–0.8 nT, which obviously determines the ultimate accuracy of the identification of the seismomagnetic effect in the test areas of Uzbekistan.

CHAPTER 5

Mesoscale variations of magnetic field in geodynamical test areas and at stationary stations of Uzbekistan

Variations of the Earth's geomagnetic field with characteristic times ranging from several months to a few years, associated with the seismotectonic processes in the Earth's crust, have been classified as transitional (mesoscale). It has been assumed that these changes are caused by the piezomagnetic effect, and that the characteristic times and dimensions are determined by the time of growth and dimensions of the earthquake focus.

The mesoscale geomagnetic field changes linked to seismotectonic processes in the Earth's crust have been investigated in the Tashkent, Fergana and Kyzyl-Kum areas by repeated route and area (eastern Fergana) surveying and stationary observations in a network of Uzbek forecast stations.

5.1 RESULTS OF MEASUREMENTS IN TASHKENT AREA

Three routes — western, eastern and transverse, 300 km long in total — were arranged in 1968 to describe the seismotectonic and geological properties of the region (Fig.21).

The western route, about 120 km long, included 18 points. It started at River Syr Darya (70 km south-west of Tashkent) and ended about Leninskiy settlement (50 km north-east of Tashkent). The first ten points were arranged in the south-western part of the route, along the Tashkent-Samarkand highway and the points 11–18 were situated in the eastern part, along the Tashkent-Chimkent highway. The route cuts across the positive and accompanying negative anomalies of the magnetic field associated with Posthercynian intrusion ($\Delta T_a = \pm 300$ nT). Its direction coincides with the active flexural-fault zone at Tashkent, and intersects the epicentre of the Tashkent eartquake of 1966 in the city itself.

The eastern route, about 90 km long passes on the eastern side of Tashkent, in the NE-SW direction, parallel to the Karzhantau fault, on the left bank of River Chirchik. It starts about Charvak, passes through Gazalkent, Yangibazar and Bektemir and ends 15 km to south-east; it consists of 18 stations.

The cross-cutting route, about 80 km long, passes north of Tashkent in the SE-NW direction, and embodies 15 stations. It begins at Sukok settlement, turns to

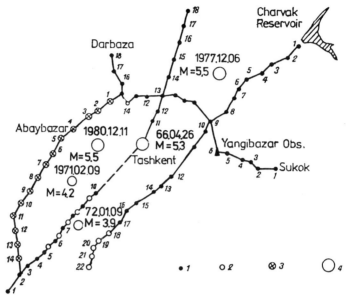

Figure 21. Layout of magnetic field measurement points in Tashkent area: (1–3) stations deployed in 1968 (1), 1970 (2) and 1975 (3); (4) earthquake epicentres.

the magnetic observatory of Yangibazar, town of Chirchik, Chernayevka settlement and Keles, and ends about Darbaza settlement upon passage through a series of Alpine faults of the Poltoratsk-Syr Darya anticlinal zone of north-east orientation.

An additional route along the Saryagach-Abaybazar-Syr Darya highway was deployed in 1975 (Fig.27). Its length was about 70 km, wherein 14 points were arranged. The route passes parallel to the western route, 15–25 km to the west, along the north-western border of the flexural-fault zone near Tashkent.

The measurements employed proton magnetometers of the PM-5, PMP, M-32, Zh-816 and TMP type. The data measured were related to the base value of the T-variometer of Yangibazar Observatory located within the area of investigation. On the average, 3 to 4 measuring cycles were conducted yearly. The magnitude of the anomalous field (ΔT_a) at selected routes was up to ±300 nT, that is the routes passed indeed over bodies having substantial magnetic parameters.

Shown in Figure 22 is the temporal behaviour of the field variations at 'calm' regular points of the routes, related to Yangibazar Observatory. It is seen that the average deviations do not exceed first units of nanotesla, which is an experimental evidence that the estimates of the measurement accuracy in our Chapter 4 have been correct. Aside from 'calm' points one also has 'disturbed' ones. The first substantial field variations were detected at a number of points of the western route. Beginning with the first measurements of 1968 at stations 5–9 one observes a gradual growth of local anomalies (Fig.23). The field variation over 2–5 years was: -15 nT at stations 5 and 9, $+13$ nT at stations 6 and 8, and $+23$ nT at

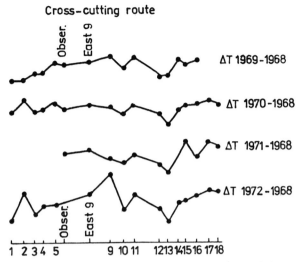

Figure 22. Normal variations of magnetic field at regular points of the cross-cutting route.

station 7 (Fig.24). These variations have a characteristic form of the field above a magnetic body. During the geomagnetic measurements from 1968 to 1971 in immediate vicinity of the southern part of the western route, on 9th February 1971 there occurred the Abaybazar earthquake ($M = 4.2$, $H = 20$ km) which could have affected the field magnitude. This is confirmed by the increase in the rate of field variation at anomalous points prior to the earthquake (Fig.23).

Repeated measurements one month after the earthquake have shown that the magnitude of the field at anomalous points changed from 5.7 to 15 nT. It is characteristic that the sign of the field anomaly changed at all points — the field grew before the earthquake and fell after the event (stations 6, 7 and 8), and vice versa (stations 5 and 9), the field behaved as if the area were 'demagnetized'.

Hence we have first detected the anomalous local change in the geomagnetic field associated with a particular incipient earthquake. The anomaly was classical; the time of its growth was 2–3 years, its maximum magnitude being about 20 nT, spatial dimension about 20 km for one sign of variations and above 30 km, with inclusion of the accompanying negative quantities at the sides of the anomaly.

Some additional earthquakes of energy class above II took place after the Abaybazar earthquake (see Fig.21). Each of them was imprinted on the magnetic field at stations of the test area. For instance, the quantity T at station 7a of the western route decreased from the end of 1970 to the beginning of 1972 by more than -25 nT (Fig.25). From January to August 1972 one observed the return of the field to the original situation. On the 9th January 1972 the Khalkabad earthquake ($M = 3.7$; $H = 5$ km) occurred. Other seismic events were not present along the route. Accordingly, the anomalous field change was associated with the growth and fall of elastic stresses at the focus of that earthquake.

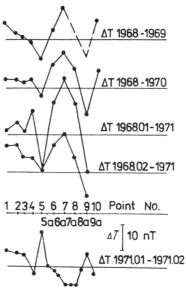

Figure 23. Anomalous variations of magnetic field along western route associated with Abaybazar earthquake on 9th February 1971 ($M = 4.2$; $H \approx 20$ km).

The magnetic field change at station 9 of the cross-cutting route is illustrated in Figure 26. The field increased slowly from 1971 and reached its maximum in the summer of 1975. From September 1975 to May 1977 it decreased sharply by 28 nT. In July 1977 the Tavaksay earthquake with $M = 5.5$ occurred some 15–20 km north-east of the station. In the first approximation, the anomalous changes can be classified in the three groups: (1) slow oscillatory growth, (2) sharp fall of anomaly, (3) change of sign of the field anomaly.

The anomalous field changes associated with the Nazarbek earthquake of 11th December 1980 were recorded at the Khumsan station. Unlike the earlier effects, the anomalies were identified by stationary measurements. In July 1980 an MPP-1 proton magnetometer was deployed there; it worked automatically, in rigorous synchronization with Yangibazar Observatory, and recorded the field magnitude every 10–20 minutes (Fig.27). During July–August 1980 an anomalous field change occurred over 15–20 days. The intensity of the anomaly was about 5–6 nT, for the accuracy of the discrimination lower than 1 nT. The distance from the measuring station to the earthquake epicentre was 60–70 km. It is noteworthy that the earthquake epicentre and the measuring station were situated in the same seismogenic zone. The effect was not observed in the direction perpendicular to the seismogenic zone, 40–45 km to the east of Yangibazar Observatory.

Hence the Tashkent area is characterized by conditions favouring studies on the seismomagnetic effect: contemporary movements of high diversification and amplitudes are combined here with strong Posthercynian intrusions having a magnetic susceptibility of 10^{-3} [cgs]. Moreover, the well developed highway system

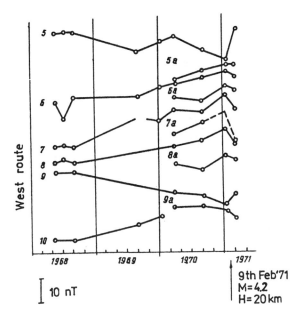

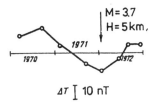

Figure 24. Anomalous magnetic field changes in time at measuring stations during incipient Abaybazar earthquake: (5–10) station numbers.

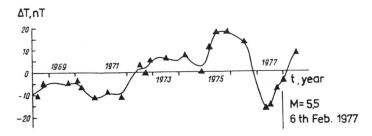

Figure 25. Anomalous magnetic effect of Khalkabad earthquake on 9th January 1972.

Figure 26. Anomalous variations of magnetic field caused by Tavaksay earthquake of 6th December 1977.

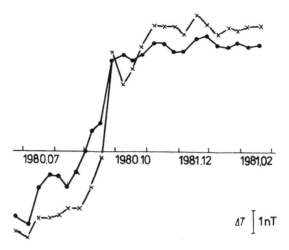

Figure 27. Magnetic effect of Nazarbek earthquake on 11th December 1980 ($M = 5.5$; $H \approx 15$ km) recorded at Khumsan Station in stationary regime. (1) ΔT for Khumsan-Yangibazar; (2) ΔT for Khumsan-Andizhan.

made simpler the field measurements. It should be emphasized that the anomalous changes in the geomagnetic field associated with the period prior to the Abaybazar, Khalkabad, Tavaksay and other earthquakes were discriminated in the western and eastern routes, which are parallel to the Poltoratsk-Syr Darya seismotectonic zone.

5.2 RESULTS OF MAGNETOMETRIC INVESTIGATIONS IN KYZYL-KUM GEODYNAMICAL AREA

The measurements in the Kyzyl-Kum area were conducted along the routes Uchkuduk-Zarafshan, Tamdybulak-Kizylkuduk, Zarafshan-Muruntau, Muruntau-Mullaly, Tamdybulak-Beshbulak and Beshbulak-Keriz. The primary part of the routes passed along the tarred highways, and it was only a section of the Muruntau-Mullaly untarred road and some points along the Tamdybulak-Beshbulak-Mullaly road that the route was unconsolidated (moving sands). All routes cut across major tectonic faults and zones of intensive fissuring or passed through tectonic discontinuities (Fig.28).

The measurements were conducted by the method of synchronous determination of the quantity T at regular stations and a reference station. The frequency of the repetition was 1 year (one cycle of measurements per year), and the measurements at each point were repeated two times a day, at different times within the measuring cycle. The distance from the reference (Tamdybulak) to the most remote point was below 80 km.

The measurements at the route points in the years 1974–1975 employed a P-5

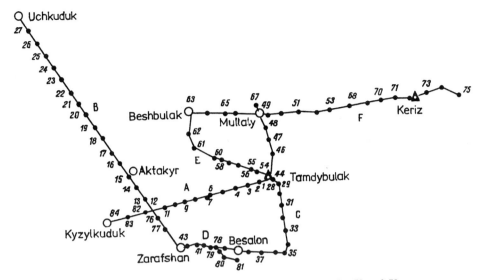

Figure 28. Layout of magnetic field measurement stations in Kyzyl-Kum area. Routes: (A) Tamdybulak-Kizylkuduk; (B) Zarafshan-Uchkuduk; (C) Mullaly-Muruntau; (D) Muruntau-Zarafshan; (E) Tamdybulak-Beshbulak; (F) Beshbulak-Keriz.

proton magnetometer, while Zh-816 magnetometers designed by Geometrics and TMP (constructed at the Institute for Geophysics of the Ural Scientific Centre of the Soviet Academy of Sciences) were used in the years 1976–1977. The standard deviation of the measurement error was 2.22 nT in 1974–1975, 1.9 nT in 1975–1976, 1.4 nT in 1974–1976 and 1974–1977, and 0.7 nT in the years 1976–1977.

The greatest field change was observed in the years 1974–1976, while the changes in the years 1976–1977 were smooth (Fig.29). This situation makes it possible to assume that the field change in the test area was anomalous in 1975. The 1975 observations were conducted in October, five to six months before the Gazli earthquakes on 8th April and 17th May 1976 (with the magnitude of 7 and 7.3).

The Kyzyl-Kum area stations were spaced 180–250 km from the epicentre of the Gazli earthquakes. Therefore the site was inside the incipient earthquake area, and in a stressed condition. This is evidenced by some other facts. During the measurements of 1974–1975, some parts of the test area exposed intensive processes of fissuring. After the earthquake the rate of fissuring fell dramatically. It appears that the Gazli earthquakes and the fissuring processes were caused by a common source. One of the anomalous field changes (by 8 nT) at the reference of Tamdybulak was identified reliably through 10-day discrete (every 10 minutes) round-the-clock measurements in the years 1974–1975, and by comparison against the data of the magnetic observatories of Yangibazar and Vannovskaya. We will discuss them in detail with reference to the field changes at the station of Tamdybulak.

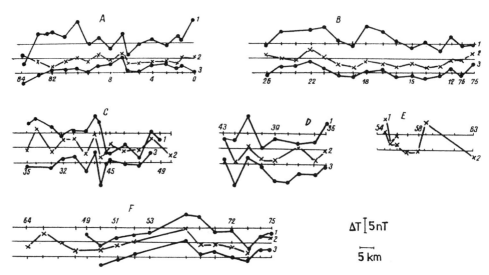

Figure 29.   Magnetic field variation over routes.   (A) Tamdybulak-Kizylkuduk; (B) Zarafshan-Uczkuduk; (C) Mullaly-Muruntau; (D) Muruntau-Zarafshan; (E) Tamdybulak-Beshbulak; (F) Beshbulak-Keriz. ΔT changes: (1) 1974–1975; (2) 1975–1976; (3) 1976–1977.

   For the most reliable discrimination of the anomalous changes in the magnetic field at the reference of Tamdybulak, statistical data processing was applied to all observations in the years 1974–1975. The discrete data obtained every 10 minutes were related to the base line of the T variometer of Yangibazar Observatory. Mean values of ΔT between Yangibazar and Tamdybulak for n = 468 data of 1974 yielded 444.3 nT, versus 436.17 nT for n = 226 in 1975. A similar difference was obtained for the data measured in 1976 and related to Yangibazar Observatory. The computed difference $\Delta T_{1975-1974} = \Delta T_{m1975} - \Delta T_{m1974}$ was −8.36 nT (index 'm' standing for mean value). This indicates that the value of T at the reference of Tamdybulak decreased by 8.36 nT during one year.

   In order to verify the anomalous field change at the reference, its magnitude was also related to regular points by statistical processing. For instance, the quantity σ_{sys} computed by the formula:

$$\sigma_{sys} = \frac{\sum \Delta T_{1975-1974}}{n} \tag{5.1}$$

was −8.007 nT; the quantity $\Delta T_{1975-1974}$ stands for the field change at individual points during the years 1974–1975, and n is the number of points.

   Hence the reliability of the determination of the anomalous field changes at the reference of Tamdybulak by 8 nT has been confirmed quite independently by another method.

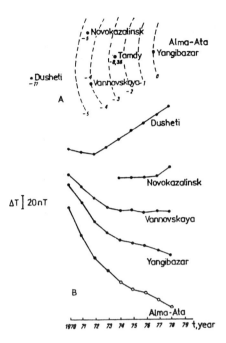

Figure 30. Anomalous change in magnetic field at reference station of Tamdybulak and distribution of magnetic-field secular course at adjacent observatories over area (A) and in time (B).

There arises a question if the field change at the above point is not caused by secular changes. The local nature of the effect is confirmed by the anomalous change related to the adjacent measuring stations. This conclusion is also supported by results of studies on the secular course in the region (Fig.30). A simple comparison shows that the field change at Tamdybulak is obviously anomalous.

5.3 RESULTS OF MEASUREMENTS IN FERGANA AREA

The Earth's magnetic field changes associated with the processes in the Earth's crust were conducted in the Fergana geodynamical area in four stages, in the years: (1) 1972–1975; (2) 1975–1977; (3) 1977–1979; and (4) 1979–1986.

First stage (1972–1973). Investigations of the variation of the geomagnetic field associated with earthquakes and other processes in the Earth's crust were initiated in 1972. They were arranged within a closed route passing along the major active deep-reaching faults and seismogenic zones (northern Fergana, southern Fergana-Kurshab, etc) (Fig.31). The spacing of the points was 4–6 km, for 750 km of the total route length. The measurements were facilitated by the proton magne-

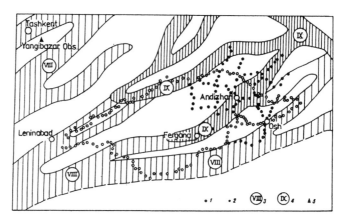

Figure 31. Layout of points of magnetic observations in Fergana area: (1–2) points deployed in 1972 (1) and 1977 (2); (3) and (4) seismogenic zones after Ibragimov (1972) for 8- and 9-degree earthquakes (3 and 4, respectively); (5) reference stations.

tometers PM-5, M-32 and PMP, and the results were related to the base line of T variometer at Yangibazar Observatory. The measurements were repeated 3–4 times per year. The accuracy of measurement of the anomalous changes was about 3 nT.

In the first period of measurements one identified a few segments with anomalous changes in the magnetic field. The most intensive of them included the anomaly in the area of Andizhan-Kampyrravat Reservoir, associated with the construction of the dam, and the anomalous change at points 1–10 (Khaidarkan-Buadil), with a density of 10–20 nT, which seems to be associated with the earthquake of the 2nd January 1974 ($K = 13$). Moreover, some other earthquakes with $K = 13$ occurred in close proximity of the routes in the years 1972–1974. In view of the low accuracy of the measurements by the apparatus applied, the inappropriate inclusion of the variations and the rare repeatability, the effects associated uniquely with earthquakes were not identified. Despite this fact, the investigations in the first years have exposed the rational potential of magnetometric studies within Fergana Valley, particularly in its southern and south-eastern parts. The statements on the activity of the south Fergana zone and its linkage of local changes to tectonic processes of considerable linear dimensions were later confirmed by the Isfara-Batken, Khaydarkan and other earthquakes with $M \geq 5$.

Second stage (1975–1977). The results of investigations of solar daily variations at individual points of the Fergana area versus Yangibazar Observatory have shown that at some places the difference between variations on disturbed days during morning hours reached up to 8–10 nT (see Chapter 4).

Therefore, since 1975 the observations have been conducted by the method of

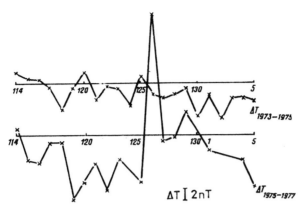

Figure 32. Anomalous changes in magnetic field along Kanibadam-Khaydarkan route associated with the Isfara-Batken earthquake of 21st January 1977.

synchronous measurements (Abdullabekov et al. 1979b). The technique involves synchronous measurements at the reference and a measuring station, with subsequent subtraction of the data. The measurements at both types of stations were facilitated by TMP magnetometers. At each point one took series of 3–5 readings at an altitude of 1–1.3 m, and the synchronization with the reference was very accurate. If micropulsations were observed or if the readings were scattered, then the number of data was increased to 7–10. The measurements at the ordinary points were fitted to reference measurements with an accuracy not lower than 1–5 s. The observations at the reference station were conducted by a compatible schedule, every 5 or 10 minutes in coincidence with measurements at the regular points. Each day before initiation of the route measurements and after the field measurements, the magnetometers were intercalibrated, and the clocks were checked against the time reference. In 1975–1977 the geomagnetic measurements were conducted at five temporary references, i.e. Yaz"yavan, Batken, Shakhimardan, Osh and Madaniyat (Fig.31).

The western part of the test area (points 1–5 and 65–130) were referred to the basic stations of Yaz"yavan and Batken while the eastern ones were related to Yaz"yavan, Shakhimardan, Osh and Madaniyat. The Isfara-Batken earthquake on 21st January 1977, with $M = 6.75$, occurred at stations 121–122, 20 km southeast of Isfara. The last observations before the earthquake were conducted in the autumn of 1975, while the next series was taken after the quake (Fig.32). The method of synchronous measurements in the years 1975–1977 had the accuracy of identification of anomalous changes of 1–2 nT.

It is seen from the drawing that the field variation along the route in the years 1973–1975 was relatively calm, as compared with the years 1975–1977. They were obviously controlled by the elastic stresses accumulated and redistributed in the zone of incipient earthquake. The jumpwise change in the field (by 10-15 nT) was recorded at points 125–126, spaced 20–25 km south-east of the epicentre. The shift can be explained in terms of a redistribution of elastic stresses after the strong shock. It should be emphasized that the epicentre of the next strong earthquake in Khaydarkan region (on 3rd July 1977) was 50–60 km east of the epicentre of the Isfara-Batken earthquake.

In the years 1973–1977 one identified anomalous field changes in the north-eastern part of Andizhan area along the Madaniyat-Aim- Kurgantepe-Karasu route; in the years 1973 to 1976 they were insignificant and did not have any directional property but were rather oscillatory (Fig.33) while in 1977 one observed anomalous decrease in the field at points 46–51. The amplitude of the maximum field change in 1977 reached 10 nT (points 46–49). The linear dimension of the identified anomaly was 30–35 km. At a distance of 10–15 km south of the route of the geomagnetic investigations in 1977 the repeated geodetic measurements along the Andizhan-Karasu railway exposed stretches with an anomalous change in the elevation of the Earth's surface.

Yarmukhamedov et al. (1979) link the contemporary fissuring and the changes in the elevation of the Earth's surface about settlements of Yuzhnyy Alamyshik and Aim and the city of Andizhan to the presence of the south Fergana flexural-fault zone, hence they point to the tectonic nature of those processes.

Third stage (1977–1979). Since 1977, together with the repeated route observations along a closed profile, repeated area surveying was also organized within Fergana Valley. In order to provide the spatial and temporal picture of field variations, about 100 additional stations were arranged in the summer of 1977 all over the territory of eastern Fergana (Fig.38). In March 1978 a repeated series of measurements was conducted at points deployed in 1973 (40–56) and at the stations of repeated area surveying deployed in 1977 (Fig.33).

Diversified anomalous changes were observed during eight months of the investigations. If one combines all stations with positive values of $\Delta T_{1978-1977}$, then it becomes possible to discriminate an area with anomalous field growth.

The points with higher differences have been situated primarily to the north of the south Fergana flexural-fault zone (see Fig.31). The latter seems to separate the area of higher values of the magnetic field from the area of lower ones.

The most considerable seismic event within the testing area was the Alay earthquake of 1st November 1978. An analysis of the spatial and temporal relationships in the geomagnetic field change over the territory of eastern Fergana (Fig.33) related to their geodetic and seismic background has made it possible to assume that the discriminated anomalous field changes have been associated with the processes preceding the Alay earthquake on the 1st November 1978. Results of magnetometric investigations confirm the nature of the contemporary movements of the Earth's crust exposed by repeated levelling. Anomalous changes in the magnetic field of one sign occur over tens of kilometres. However, the intensity of the anomalies inside the area varies disorderly and in jumps, and local subareas with a gradient of changes of 10–14 nT are encountered. One cannot observe smooth and isometric changes. The relation of the anomalous changes in the magnetic field to the Alay earthquake is also evidenced by the fact that similar effects were not observed in the magnetic field 2–3 years after the shock, and seismic events with $K \geq 10$ were not distinguished.

Fourth stage (1979–1986). Detailed investigations of the geomagnetic field changes associated with processes in the Earth's crust were initiated in eastern Fergana in 1979 (Abullabekov et al. 1981). The studies were carried out on a contract basis in co-operation with the Institute of High Temperatures of the Soviet Academy of Sciences. The measurements involved a combination of repeated area surveying and stationary observations (Fig.34).

Since long-term measurements of the geomagnetic field changes seem to have been conducted first at such a large number of stations and with that high frequency then it is purposeful to discuss them more thoroughly (see Fig.35).

Unlike the results obtained in the earlier years (1975–1978), one did not detect any serious changes in the field in the years 1979–1985, and no trend was observed in slow changes. The maximum variations in individual cases were up to 8–10 nT. The diversification of field changes over the area varied from place to place, so that synchronous changes were noted in some cases in a group of points, and different phases were observed in other instances. General regularities typical for all stations of the test area were impossible to expose. Nonetheless, the magnitude was not constant at any point.

The eastern Fergana territory area was seismically calm during the investigations. There were no records of earthquakes with $K \geq 10$–11. However, some ten strong shocks with $M = 4.5$–6.5 took place around the test area at a distance $R \leq e^M$. The field changes in such conditions were insignificant and chaotic. Before the strong seismic events the variation of the field was clearly ordered (Fig.36b).

As seen in the drawing, some anomalous changes with an intensity from units to 12–13 nT have been identified in the repeated area surveying in eastern Fergana before the remote seismic events. The earthquake epicentres were situated in different directions from the test area: to the south in the case of the Alay earthquake of 1st November 1978, to the west for the Chimion earthquake of 6th May 1982, to the south for the Afghan earthquake of 30th December 1982, to the east for the Chinese earthquakes of 13th February and 5th April 1983, and to the west for the Pap earthquake of 1984. However, despite the various directions of seismic events, the anomalous changes have some common characteristics. The maximum variations at individual points and the contour zones with identical field changes in all cases are attributed to the same parts of the test area (Fig.36). The dimensions of the contour zones are comparable with dimensions of the Earth's crust blocks identified by various authors using different methods (Sadovskiy 1979 and others). It is possible that the slow anomalous changes appear as the result of

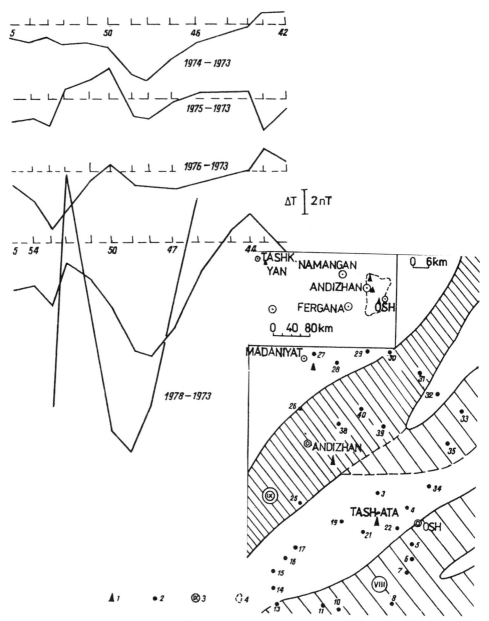

Figure 33. Anomalous variation of magnetic field along Madaniyat-Karasu profile due to Alay earthquake of 1st November 1978 (top).

Figure 34. Layout of fine-measurement stations for the magnetic field changes in eastern Fergana: (1) stationary points; (2) seismogenic zones (after Ibragimov 1978) for 8- and 9-degree earthquakes; (3) borders of test area; (4) points of repeated area measurements.

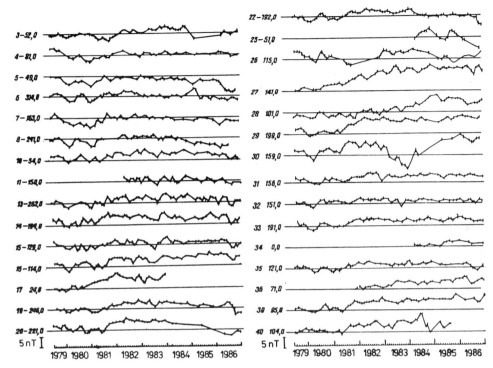

Figure 35. Magnetic field changes at regular points of eastern Fergana test area in the years 1979–1986.

the interaction of the Earth's crust blocks and the redistribution of elastic stresses between them.

The linkage of the maximum changes to definite parts of the test area bears witness of the non-uniform spatial distribution of elastic stresses. The points seem to be differentiated as sensitive and less sensitive or those which carry more and less information. The highly informative points are a kind of indicators and are most valuable in organization of routine forecast studies in geodynamical test areas.

A more critical analysis of the results obtained in the test area shows that repeated area surveying in confined territories cannot provide a complete picture of spatial field changes. Repeated surveying must be conducted on larger territories. The frequency of repetition should be based on results of stationary observations.

Mesoscale anomalous changes in the magnetic field have also been studied on the basis of the data obtained from stationary measurements. Mean monthly differences of the magnetic field between couples of close stations were constructed for Madaniyat vs. Tash-Ata; Madaniyat vs. Andizhan and Andizhan vs. Tash-Ata. Before a violent earthquake occurring within the radius $R \leq r^M$ km one

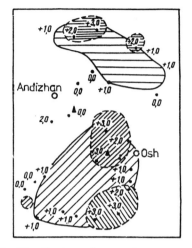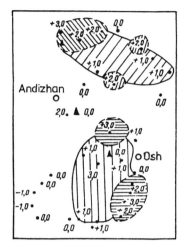

Figure 36. Distribution of anomalous changes in magnetic field over the test area before Chimion (a) and Chinese (b) earthquakes. (a) March 1982 – August 1981; (b) = October 1982 – August 1981.

Table 8. Mesoscale magnetic effects of violent earthquakes in Uzbekistan.

Date	Co-ordinates		K	M	H_{km}	Dnsty	Durn	Lin.	Dist.	Notes
	ϕ	λ				nT	day	dim.	km	
9 Feb 71	41.25	69	12	4.2	5-20	± 3-15	30	30-40	0-30	11pts
8 Apr 76						-4-5/yr			180-200	Tblk
31 Jan77			14.2	5.75	20	+18-7	510	60-70	0-40	13pts
6 Dec 77			14	5.5	15-20	-35	375		15-20	1 pt
1 Feb 78	39.4	72.6	16	6.8	20-30	-10	1100	50-60	100-150	10pts:
						+15				K-Md
11 Dec 80	41.3	69.2	13.4	5.2	10	7	150		90	Khum
6 May 82	40.2	71.5	14.4	5.8	15-25	-4	30		100	And
						-3-4	210		102	TaAt
						-3-4	435		122	Mad
						+3-4	450			Chim
										Reg:
						± 2-4		40-50	50-120	E Fer
13 Feb 83				6.1		5	365		280	Mad
						8	510		280	And
						8	510		280	TaAt
						± 2-4		40-50	250-350	Reg:
										E Fer
17 Feb 84	40.9	71.7	14	5.6		4	200		75	Char
23 Mar 84				6.8	25-35					Shu

(Dist) Distance to the point of measurement, km; (11pts) effect pronounced at 11 points (stations); (Reg) Regular points; (Tblk) Tamdybulak; (Kar) Karasu; (Mad) Madaniyat; (K-Md) Karasu - Madaniyat; Khum = Khumsan; (And) Andizhan; (TaAt) Tash-Ata; (Chim) Chimion; (Fer) Fergana; (Char) Chartak; (Shu) Shurchi.

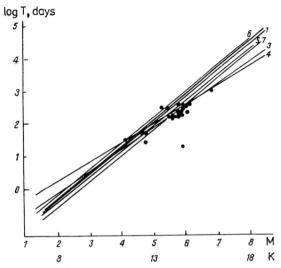

Figure 37. Relationship between duration of anomalous mesoscale changes in magnetic field and the magnitude and class of earthquake. (1) Myachkin et al. 1975; (2) Sidorin 1979; (3) Whitcomb et al. 1973; (4) Rikitake 1979; (5) Sholz et al. 1973; (6) and (7) Tsubokawa 1969 and 1973. Points denote experimental data measured in Uzbekistan.

noted anomalous field changes at regular and stationary points of the test area (Abdullabekov et al. 1986).

The mesoscale magnetic anomalies are observed a few months before crustal earthquakes. The intensity of the anomalous changes between couples of adjacent stations varies from first units to 5–6 nT. For instance, the field change before the Chimion earthquake on the 6th May 1982 ($M = 5.8$ and $R = 100$ km) began 5 months before the earthquake, had an intensity of 1.5–2 nT, and lasted 6–7 months. The change of sign occurred 2–3 months before, and the field returned to the original level thereafter.

The highest intensity and duration of the anomalous changes were observed during two strong Chinese earthquakes on the 13th February and 5th April 1983 ($M = 6.0$–6.1 and $R = 280$–300 km). The intensity of the anomaly ranged from +3 to -2 nT, for the total duration of 10–11 months. The field increased before the earthquakes and was decreasing after the shocks.

Subsequent anomalous changes were noted before the Maylisay earthquake on the 5th November 1983 ($M = 4.5$; $R = 60$ km), Pap earthquake on the 17th February 1984 ($M = 5.6$; $R = 120$ km) and Dzhirgital earthquake on 27th October 1984 ($M = 6.3$, $R = 200$ km). The parameters of these and other magnetic anomalies identified in Uzbekistan test areas and at stationary forecast stations are given in Table 8.

Linear dimensions of the anomalous effects were determined in five cases. They are in the range from 30–40 to 60–70 km. In other cases the distances from the epicentre to a point of observation are given. The linear dimensions of the effects are however difficult to provide by this data. It is more reliable to determine them by data of repeated route and area surveying. The characteristic times are more complete and valuable if found from stationary measurements.

The characteristic dimensions discriminated as the result of the studies have been compared with the dimensions of the Earth's crust blocks singled out by many investigators (Sadovskiy 1979, Chigarev 1980, Vilkovich et al. 1974, and others).

The distances from the epicentre to measuring stations vary from tens to first hundreds of kilometres. The changes at distances exceeding 100 km seem to be caused by the larger (regional) rank processes. Unfortunately, the existing network of dense regular points in separate test areas (Tashkent, Fergana and Kyzyl-Kum), and the rare network of stationary forecast stations do not permit so far the regional description of field changes. However, some facts are very promising. The effects of the Alay earthquake of 1st November 1978 have been recorded at distances of 100–150 km to the north of the epicentre (Mavlanov et al. 1979a) and 130–150 km south of it (Skovorodkin et al. 1980). The anomalous changes associated with the Isfara-Batken earthquake on the 31st January 1977 have been identified in Fergana (Abdullabekov et al. 1979) and Garm (Skovorodkin et al. 1983) test areas. The spacing of the segments of pronouncement of anomalous changes is 150–200 km.

The anomalous effects of the Gazli earthquakes of 1976 in the Kyzyl-Kum area (180–250 km), of the Chinese earthquake of 13th February 1986 in eastern Fergana (280 km) and of the Gazli earthquake of 20th March 1984 at stations of Shurchi, Samarkand and Khavatag (200–300 km) provide evidence on the existence of anomalous effects of the subseguent higher rank measurable in first hundreds of kilometres. The time of variation of geomagnetic field before the earthquakes varied from tens to several hundred days. One also noted a clear relationship between the energy of the earthquakes and the time of appearance of the effect (Fig.37).

The stronger the earthquake the longer is the duration of the magnetic effect. This finding has been confirmed by numerous researchers (Myachkin et al. 1975, Rikitaki 1979, Sidorin 1979, Zubkov & Migunov 1977, Sholz et al. 1973, Tsubokawa 1969, 1973, and Whitcomb 1973). The characteristic time of the effects cannot however be always determined from the temporal picture of field variations. For instance, in the cases of Abaybazar and Tavaksay earthquakes (Figs 30, 32), the effect was pronounced as a trend much earlier than the above relationships. Nonetheless, from a certain time onward, the gradient of the anomaly or a sign change are subject to a dramatic transformation. The instant of sign change or jumpwise variation of the gradient can be regarded as an origin of the anomalous precursory effect of earthquake. The anomaly was irreversibly jumpwise before the Nazarbek earthquake (Fig.33). In the third case the anomaly was displayed as an embayment (Fig.31).

CHAPTER 6

Fast changes in the Earth's magnetic field

This type of variations includes the changes in the magnetic field of the Earth with characteristic times from hours and days to weeks, associated with processes in the Earth's crust. The existence of such variations is widely covered in bibliographical sources (Lapina 1953, Shapiro 1966, Abdullabekov & Maksudov 1975, Golovkov et al. 1977, Skovorodkin 1980, and others).

The fast changes in the Earth's magnetic field were measured in Uzbekistan test areas by the method of stationary observations in a network of absolute proton magnetometers. At first the stationary merasurements were conducted at epicentres of heavy earthquakes (Alay 1974, Gazli on 8th April and 17th May 1976, Isfara-Batken 1977, Tavaksay 1977 etc); the observations were visual (Chapter 7). The measurements were conducted every 5–10 minutes, synchronously with Yangibazar Observatory. One used TMP magnetometers designated for field route surveying. The stationary routine observations on the territory of Uzbekistan have been conducted since 1978. With deployment of new types of stationary proton magnetometers — PM-001, APM, MPP-1 and MPP-1M — the network of the stationary observations was gradually expanded (Fig.38). The stations were located primarily in seismogenic segments or in their immediate vicinity: stations Khumsan and Lenin-Yuly along the Karzhantau fault and Poltoratsk-Syr Darya flexural-fault zone, four stations at Charvak Reservoir along the Khumsan-Yusupkhana profile; stations Pishkoran, Nanay and Sarychelek along the ·active north Fergana fault; stations Tash-Ata, Andizhan, and Madaniyat across the series of south Fergana faults, and the Chimion station close to the south Fergana fault.

Figure 39 illustrates the working schedule of the network of stationary magnetic stations. One station operated from 1968 to 1978 (magnetic Yangibazar Observatory), and the number of stations increased gradually from 1978 to exceed 20 at the end of 1984.

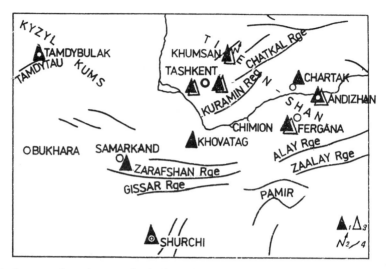

Figure 38. Layout of stationary physical stations in Uzbekistan: (1) magnetometric stations; (2) dipole electric probing (DEP) station; (3) pulsed electromagnetic field stations; (4) mountain ridges.

6.1 RESULTS OF STATIONARY MEASUREMENTS OF FAST CHANGES IN GEOMAGNETIC FIELD

Variations of the geomagnetic field at the above stations were characterized by some general features of the territory and individual peculiarities of each station. The intensity of the daily variation of the field itself reaches tens of nanoteslas and slightly exceeds the magnitude of the effects observed in the test areas. It is therefore convenient to take the difference of field variations between couples of stations. In such an approach, the field of external sources is eliminated. If one operates with the differences of synchronous measurements one also uses a certain spatial filter and makes it possible to identify local effects of crustal sources.

In the first stage of investigations the results of the stationary geomagnetic measurements in all test areas were presented as difference graphs for each station related to Yangibazar Observatory. The field differences have been characterized by a spectrum of mesoscale (transitional) and fast field changes. The spectrum was very wide as it ranged from several hours and days to several months and even longer spans of time. The amplitude of the field changes also varied from fractions and units to tens of nonotesla.

Beginning with the temporal parameters of the investigated processes in the Earth's crust one determined a wide spectrum of magnetic field differences related to Yangibazar Observatory. Average values for days, three days, five days, fifteen days, a month, a year etc were found.

Station	Coordinates		Type of instr.	Years 19..						Notes
	ϕ	λ		80	81	82	83	84	85	
Andizhan	40°75	72°35	MPP-1	————————————————						IS Uz
										IVTAN
Madaniyat	41°00	72°45		————————————————						ibid.
Tash-Ata	40°60	72°55		————————————————						
Yangibazar	41°25	69°70		————————————						KOMEIPZ
										UzAN
Tamdybulak	41°35	64°65		— —						ibid.
Khumsan	41°75	70°10		————————————						ibid.
Lenin Yuly	41°40	79°10		—————————						ibid.
Chimion	40°25	71°50		————————						ibid.
Khavatag	40°00	69°05		—————						
Shurchi	37°90	67°30	PM-001	————						
Chartak	41°20	71°90	MPP-1	————						
Dzhizak	40¹0	67°70		——						
Urgut	39°40	67°20		——						
Nanay	41°60	71°70	APM	————————————————						IS Uz
										IZMIRAN
Pishkoran	41°30	71°30		————————————						
Charvak	41°70	70°10		————————————						
Baka Shel	41°65	70°35		————————————						
Kan	40°20	71°40		————————————						
Batken	40°10	70°70		————————————						
Karakyr	40°50	63°40		—						IS Uz
										IFZAN
										IS Kaz
Dzhingildy	41°10	63°30	MPP-1	—						
Tsvetushchiy	40°45	63°15	MPP-1	—						KOMEIPZ

Figure 39. Working schedule of stationary magnetic stations in Uzbekistan test areas. (IS Uz) Institute for Seismology of the Uzbek Academy of Sciences; (IVTAN) Institute of High Temperatures of the Soviet Academy of Sciences; (KOMEIPZ) Scientific Committee of the Institute for Seismology of the Uzbek Academy of Sciences; (IFZAN) Institute for Earth Sciences of the Soviet Academy of Sciences; (IS Kaz) Institute for Seismology of the Kazakh Academy of Sciences.

The mean daily values of the field have been computed by averaging the every-hour or every-twenty minute field differences over days, while the subsequent field differences (for three days, fifteen days, etc) were obtained by simple sliding averaging of mean daily field values.

The next operation consisted in the separation of anomalous field changes from

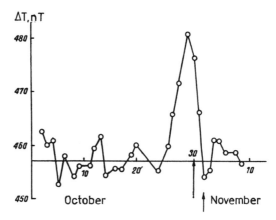

Figure 40. Short-term magnetic effects of Alay earthquake with M = 6.5 on 1st November 1978 at Andizhan station.

their background. The effects exceeding the determined standard deviation were processed in the discrimination of anomalous variations.

More details will be discussed for the most interesting and characteristic magnetic effects displayed in the test areas.

The first stationary studies on the dynamics of geomagnetic field were initiated at Andizhan. Routine measurements were conducted from October 1978 at a point located 7 km south of the city of Andizhan. TMP magnetometers were used two times a day (10:00 and 16:00 local time) and later, from 1st December 1978, five times a day (10:00, 12:00, 15:00, 16:00, 17:40); the measurements were synchronous with Yangibazar Observatory (Fig.40). The variations of the mean daily field difference Andizhan-Yangibazar found from the above series were characterized by irregular oscillations with an amplitude of ±2.5–3.0 nT.

On the background of the above oscillations, which reflected the inhomogeneities in the course of geomagnetic variations, we discriminated an intensive anomaly ΔT generated on 26th-27th October, which reached its maximum (23 nT) on 30th October; the field difference sharply fell between 31st October and 1st November, and on 2nd November 1978 returned to its initial level.

Comparison of the field magnitude at the observatories of Yangibazar and Charvak has shown that the difference between the above points remained constant during the entire period. From this it follows that the source of the anomalous variation was closer to Andizhan. From the data of Yangibazar Observatory, the last ten days of October were characterized by an undisturbed magnetic field. On the basis of the anomalous changes in the magnetic field, combined with hydrogeoseismological and other data we predicted the approaching violent earthquake (Mavlanov et al. 1979, Shapiro & Abdullabekov 1982). This referred

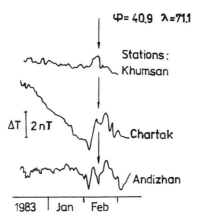

Figure 41. Short-term magnetic anomalies of Pap earthquake on 17th February 1984 in a network of stationary stations.

to the violent earthquake of M = 6.8 on 2nd November 1978 at 1:43 a.m. local time (or 1st November at 19:43 Greenwich time), at a distance of 120 km south of Andizhan in Alay Valley.

Studies on the dynamics of the geomagnetic field in the Fergana area permitted the magnetologists to predict reliably the time and approximate place of the violent Alay earthquake of 1st November 1978; this prediction was first in the USSR and embodied a number of other forecast methods. In a few subsequent years a seismic calm was observed in the territory. Strong seismic events did not occur in Fergana Valley until 1982.

Figure 41 illustrates the anomalous changes in the magnetic field at stations Khumsan, Chartak and Andizhan associated with the Pap earthquake on 17th February 1984. The epicentre of the earthquake was in the southern Fergana seismogenic zone some 100 km from Khumsan station, 80 km from Andizhan station and 25–30 km from Chartak station. The graphs show mean daily differences of the magnetic field between the stations and Yangibazar Observatory. The characteristic time varied from 15 days (Khumsan station) to one month (stations Chartak and Andizhan). The intensity of the anomaly in the most remote station of Khumsan was 1 nT, versus about 1.5 nT for Andizhan and 2 nT at Chartak. The anomalies were positive. The origin of the phenomenon occurred 8–10 days before the quake. The anomalous changes were not present at the stations spaced more than 150 km from the epicentre.

Figure 42 shows the anomalous changes in the magnetic field identified at stations Madaniyat (a), Andizhan (b), and Tash-Ata (c), related to Yangibazar Observatory, before the close Markhamat earthquake on 26th February 1986. The

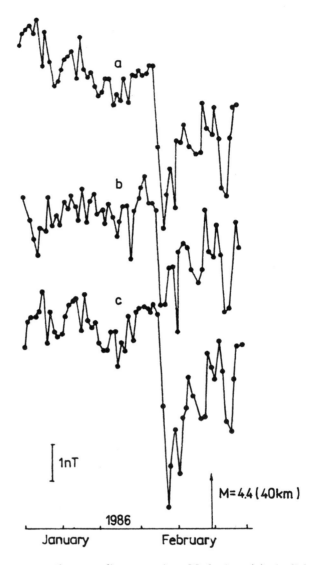

Figure 42. Short-term magnetic anomalies at stations Madaniyat (a), Andizhan (b) and Tash-Ata (c) before the Markhamat earthquake of 26th February 1986.

epicentral distances were respectively 15, 40 and 60 km. The effect was pronounced some 15–25 days before the quake. The intensity of the anomalies at the close station of Tash-Ata was 5 nT, while that at the remoter stations of Andizhan and Madaniyat was 3–4 nT. The gradient of the incipient anomaly was jumpwise. The effect was pronounced only at eastern Fergana stations. It was not observed

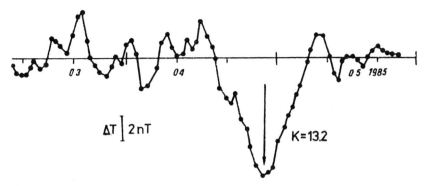

Figure 43. Characteristic bay-type magnetic field anomalies prior to close violent earthquake.

at other stations of Fergana Valley situated further than 120 km (Chimion and Chartak).

In order to analyse more closely the short-term magnetic effects we constructed temporal series of mean daily differences of the magnetic field at all stationary stations and Yangibazar Observatory. Catalogues of the earthquakes within the radius $R \leq e^M$ km (Dobrovolskiy et al. 1980) or $R \leq 30P$ (Ulomov 1978) were used and mutually compared. One observed numerous magnetic anomalies before seismic events.

Table 9 contains the basic information on the earthquakes and the recorded magnetic effects, the duration and intensity, the characteristic times before the event, times of sign change in the anomalous field, etc. The earthquake parameters are recorded in the yearly catalogues *Earthquakes in the USSR* (1978–1982), *Earthquakes in Middle Asia and Kazakhstan* (1978–1982) and operational bulletins of the Tadzhik and Uzbek Academies of Sciences (TISSS AN Tadzhik Academy of Sciences and KOMEIPZ IS Uzbek Academy of Sciences).

The analysis has shown that in 70–80% of cases the anomalous changes were accompanied by earthquakes, while in 20–30% of events anomalous changes in the geomagnetic field were not present. The shape of the anomalous changes was primarily bay-type, with the negative sign. The intensity varied from first units to 8–10 nT, or up to 20 nT and even more in some particular cases. The most typical bay-type anomaly of the magnetic field at Khavatak station is illustrated in Figure 43; it was caused by an earthquake of $K = 13.2$ some 100 km away in southern Tien-Shan.

Table 9. Catalogue of short-term magnetic anomalies discovered at stationary stations of Uzbekistan.

Date	Green. time	Coordinates ϕ	λ	K	M	H	P_{30}	Dnsty nT	Dur day	T_1 day	T_2 day	R, km	S
1	2	3	4	5	6	7	8	9	10	11	12	13	14
						1978							
3 Oct	14.20	39.4	74.7	15	6.1	33+ 5	525	−7	11	6	3	240	A
1 Nov	19.43	39.4	72.6	16	6.3	20-30	1050	+23	9	6	5	150	
3 Nov	00.57	39.4	72.47	14	5.2	15-20	269	−4	4	3	2	150	
17 Nov	12.59	33.55	70.54	13	5.0	20	132	to −14	10	6	2	300	
						1979							
20 Jun	11.50	36.5	70.5	14	5.6	200	269	−7 +3	5	3	3	500	
26 Jun	03.00	36.6	71.2	14	5.6	230	269	−7 +3	11	9	4	500	
20 Mar	03.50	36.5	70.0	15	6.1	230	525	−5	15	3	5	500	
25 Nov	06.03	41.55	71.3	9	2.3	10	10	−3.5	15	3	4	10	
03 Dec	21.36	41.43	71.9	10	3.3	5	13	−5	15	3	5	30	
03 Oct	21.36	41.43	71.9	10	3.3	5	13	−5	15	3	5	30	N
						1980							
13 Jan	05.54	39.4	72.3	13.4	5.2		190	−2.5-3	20	10	10	160	A
03 Mar	10.59	40.9	71.3	11	-	5	33	6-7	20	20	13	50	
14 Mar	07.05	41.0	71.3	9.6	3.1		15	6	20	20	-	50	
23 Mar	03.46	41.0	72.0	10.3	3.5		20	−2.5	25	25	21	50	
11 Jul	11.47	40.2	70.4	13.5	5.3		200	−6+13	35	23	7	100	C
11 Jul	11.47	40.2	70.4	13.5	5.3		200	2.5+3.5	25	13	13	190	A
11 Jul	11.47	40.2	70.4	13.5	5.3		200	−3	22	17	14	160	K
20 Jul	01.32	41.5	72.4	11.7	4.3		66	3	10	13	5	63	N
29 Mar	09.10	41.2	71.7	10.2	3.4		13	−2.5+3	31	23	10	30	A
						1981							
02 May	16.04	36.6	71.0	16.0	6.7	210	1050	−2+2.5	17	11	6	470	A
02 May	16.04	36.6	71.0	16.0	6.7	210	1050	−2	22	3	0	560	K
02 May	16.04	36.6	71.0	16.0	6.7	210	1050	−3	29	27	13	420	C
02 May	16.04	36.6	71.0	16.0	6.7	210	1050	+3	27	22	14	470	M
02 May	16.04	36.6	71.0	16.0	6.7	210	1050	4-5	30	25	17	470	T
30 May	15.67	40.9	72.3	9.6	3.1		13	−3+3.5	22	14		19	A

6.2 ANALYSIS AND DISCUSSION OF RESULTS OF STATIONARY MAGNETIC FIELD OBSERVATIONS

The distribution of the intensity of anomalous magnetic field effects related to the energy of earthquakes indicates that there is no explicit relationship (Fig.44). The magnitude of the changes (both positive and negative) in most cases varies from units to 8 nT, and it was only in five cases out of 77 that it ranged from 14 to

Table 9 (continued).

Date	Green. time	Coordinates φ	λ	K	M	H	P$_{30}$	Dnsty nT	Dur day	T$_1$ day	T$_2$ day	R, km	S
1	2	3	4	5	6	7	8	9	10	11	12	13	14
							1982						
02 Jan	17.44	41.0	72.5	9.3	3.2		13	4	5	5	4	5	M
02 Jan	17.44	41.0	72.5	9.3	3.2		13	−2	3	5	2	30	A
07 Mar	12.24	33.1	72.5	14.0	5.6	140	269	−4	11-12	7	0	255	C
06 May	15.42	40.3	71.3	14.4	5.3	15-25	371	−6	21	7		10	C
							1983						
13 Feb	01.40	40.0	75.2	15.6	6.7		340	−3-4	10	9	3	270	A
13 Feb	01.40	40.0	75.2	15.6	6.7		340	−2	30	13	1	500	K
13 Feb	01.40	40.0	75.2	15.6	6.7		340	3	25	17	3	246	T
13 Feb	01.40	40.0	75.2	15.6	6.7		340	−2	11	9	5	350	C
05 Apr	06.56	40.1	75.4	14.3	6.0		500	−1.5-2	15	13	6	350	C
05 Apr	06.56	40.1	75.4	14.3	6.0		500	1-1.5	20	3		50	K
12 Sep	15.42	36.7	70.9	15.5	6.1	200	700	−1.5+2	10	10	5	500	H
12 Sep	15.42	36.7	70.9	15.5	6.1	200	700	−3--4	21	12	9	330	C
12 Sep	15.42	36.7	70.9	15.5	6.1	200	700	+3	10	10	6	405	G
06 Oct	10.01	41.1	72.6	10.0	3.3		13	−2	15	5	5	10-15	M
03 Dec	00.11	39.9	69.1	10.0	3.3		13	−1.5+2	30	25	15	15	K
16 Dec	13.15	39.4	73.4	73.9	5.9		450	−2	30	30		240	H
16 Dec	13.15	39.4	73.4	73.4	5.9		450	−1.5	12	10	3	130	A
16 Dec	13.15	39.4	73.4	14.7	5.9		450	−3	20	9	6	400	K
16 Dec	13.15	39.4	73.4	14.7	5.9		450	−2	4	3	2	200	C
30 Dec	23.15	36.5	70.6	16.7	7.0	210	1600	−6	16	10		310	S

23 nT. In this case the above relationship was analysed without inclusion of the epicentre distance and the duration of the effect.

The duration of the effects varies from several to 30–40 days. The graphical representation in Fig.45 shows that there is a wide spectrum of the duration of fast magnetic effects prior to earthquakes. One can make a very careful conclusion that a tendency toward longer duration of the effects is pronounced with increasing energy of the earthquake. This is better visible for the subcrustal Hindu Kush earthquakes.

The intensity of the effect is independent of the distance and lies in the range of ±8 nT , with the exception of the anomalous changes before single earthquakes, which reach 14–23 nT. In this case one cannot talk about any relationship between the intensity and the epicentre distance whatsoever.

The dependence of the intensity of an effect on distance was investigated with inclusion of the earthquake energy. The intensity was related to the conditional

Table 9 (continued).

Date	Green. time	Coordinates φ	λ	K	M	H	P30	Dnsty nT	Dur day	T1 day	T2 day	R, km	S
1	2	3	4	5	6	7	8	9	10	11	12	13	14
							1984						
27 Jan	23.01	36.3	70.3	15.2	6.1	160	600	+2	13	12	5	310	S
01 Feb		36.3	70.1	15.2	6.1		600	−4	5	2		500	G
12 Feb		41.0	71.1	11.5	4.4		76	−2.5	30	15	12	75	H
13 Feb	10.07	40.0	63.3	11.1	4.0		45	−2.5+3	20	10		100	G
16 Feb	17.13	36.6	70.7	15.6	6.6	200	900	2	10	6	3	500	A
16 Feb	17.13	36.6	70.7	15.6	6.6	200	900	−1.5+2	13	3	5	400	C
16 Feb	17.13	36.6	70.7	15.6	6.6	200	300-900	−3.5-5	13	5	1	320	S
16 Feb	17.13	36.6	70.7	15.6	6.6	200	300-900	−2.5-3	20	14	11	400	G
17 Feb	23.26	40.9	71.7	14.0	5.6		275	−2-2.5	14	13	10	115	A
17 Feb	23.26	40.9	71.7	14.0	5.6		275	−2-2.5	13	9	6	90	C
19 Mar	20.23	40.4	63.2	17.0	7.2		2100	+4	23	20	13	130	D
								− 2.5	40	27	26	525	G
								−4	30	23	16	500	S
								−4-5	25	24	9	200	C
								−3	30	27	13	330	A
19 Apr	02.53	36.6	70.3	14.2	5.7	130	450	−2.5-3	20	10	7	350	S
01 Jul	10.12	36.6	70.3	14.3	6.0	200	500	3	16	16	9	320	S
13 Jul	01.40	40.3	72.5	11.5	4.2		75	1--1.5	20	10	5	15	A
14 Mar	11.46	40.4	63.5	13.5	5.3		200	No explicit anomaly				130	D
22 Mar	13.01	36.2	70.3	14.4	5.3	150	350	3	14	11	7	320	S
26 Oct	20.22	39.2	71.2	14.7	5.9		450	3	25	20	14	220	G
26 Oct	20.22	39.2	71.2	14.7	5.9		450	4	11	5	3	340	S
26 Oct	20.22	39.2	71.2	14.7	5.9		450	2	45	40	20	120	C
26 Oct	20.22	39.2	71.2	14.7	5.9		450	−2-2.5	44	22	16	195	T
27 Apr	01.31	33.9	72.6	13.4	5.2		190	1.5-2	9	3	4	190	T
27 Apr	01.31	33.9	72.6	13.4	5.2		190	1.5-2	9	3	4	120	M
13 Oct	15.59	40.25	69.3	14.9	6.0		500	No explicit effect				230	A
13 Oct	15.59	40.25	69.3	14.9	6.0		500	1.5-2	15	7	5		T
13 Oct	15.59	40.25	69.3	14.9	6.0		500	1.5-2	17	3	6		M

(S) stations: (A) Andizhan, (N) Nanay, (C) Chimion, (K) Khumsan, (M) Madaniyat, (T) Tash-Ata, (H) Chartak, (G) Khavatag, (S) Shurchi, (D) Tamdybulak; (P_{30}) 30 focus dimensions, after Ulomov (1970); (R) distance from earthquake epicentre to measuring station; (T_1) time to shock, days; (T_2) time to sign change.

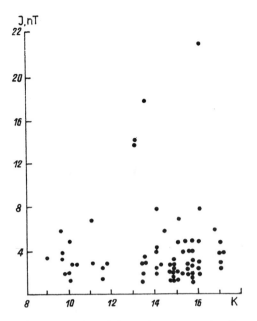

Figure 44. Dependence of the intensity of short-time magnetic effects on earthquake class.

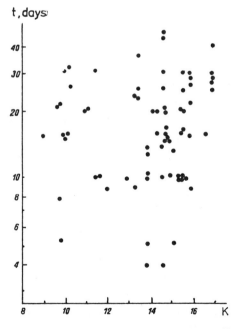

Figure 45. Dependence of the duration of short-term magnetic effects on earthquake class.

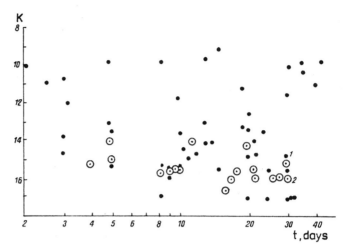

Figure 46. Distribution of sign change in short-term magnetic effects as a function of earthquake class: (1) crustal; (2) subcrustal.

relative seismic energy $\frac{\log E}{\pi R^2}$ in which R = distance from the point of recording to earthquake epicentre and πR^2 = area of circle. Like in the preceding examples it is difficult to identify any clear relationship between the parameters analysed. The short-term effect of the magnetic field is obviously independent of the distance and energy of earthquakes.

In most cases the anomalous changes in the magnetic field are bay-type and reversible. As a rule, the earthquake occurs after the sign of the magnetic effect changes, with a certain dependence on the energy class. The time of the sign change varies from 0 to 20 days. There is no clear-cut relationship between the earthquake energy and the time of sign change of the anomalous field (Fig.46).

It is seen from the above analysis that the spectrum of characteristic times of 'fast' magnetic effects is contained in a fairly narrow band stretching from hours and days to first tens of days. At first sight, for operational forecast it is sufficient to record the second type of variations at several stations, so that the earthquake could be predicted after the sign changed. However, the lack of a relationship between the intensity of short-term effects and the energy class of earthquakes, or any relation to the distance, makes difficult the forecast of the strength and place of expected shocks. It is possible that, at the present stage of research and recording of the fast effects, the problem can be solved only for the prediction of the date of earthquake. The effectiveness is a decisive factor in this case.

Despite the lack of a correlation between the parameters of the short-term magnetic effects (intensity, duration and linear dimension) versus earthquake parameters (class, magnitude, focus depth, etc) one can note a few regularities which provide a basis for optimization of the monitoring network, elaboration of proper techniques, etc.

The density of fast anomalous effects in the magnetic field varies from units to tens (and more) of nanoteslas. The form of the anomalies is bay-type, mostly with the negative sign. The characteristic time varies from several days to 2–3 weeks.

The stretch of the effect depends on earthquake energy — the stronger the seismic event the larger is the distance and the greater the amount of the stations where the anomaly is recorded.

In combination with magnetometric measurements at many stations we conducted investigations on the variation of the real pulsed electromagnetic field of the Earth (Chapter 8). The joint analysis of the results of the studies on the variation of the magnetic and pulsed electromagnetic fields shows that their characteristic times are similar.

CHAPTER 7

Anomalous variations of the Earth's magnetic field at epicentres of violent earthquakes

Routine magnetometric studies were conducted at epicentres of the past earthquakes: Alay on 11th August 1974 ($M = 6.8$), Gazli on 8th April ($M = 7.0$), 17th May 1976 ($M = 7.3$) and 19th March 1984 ($M = 7.1$), Isfara-Batken on 31st January 1977 ($M = 6.4$), Tavaksay on 6th December 1977 ($M = 5.5$), Nazarbek on 11th December 1980 ($M = 5.5$), Chimion on 6th May 1982 ($M = 5.8$), Pap on 18th December 1984 ($M = 5.6$), etc (Shapiro & Abdullabekov 1977, Abdullabekov et al. 1984, Mavlanov et al. 1983, Abdullabekov 1983, Shapiro & Abdullabekov 1978a, and others).

7.1 EPICENTRE OF GAZLI EARTHQUAKES OF 1976 AND 1984

The studies on the variations of the Earth's geomagnetic field at the epicentre of the Gazli earthquake of 1974 were conducted with a highly sensitive proton magnetometer (TMP-type, having the sensitivity of 0.1 nT), designed at the Institute for Geophysics of the Soviet Academy of Sciences (UNTs AN SSSR). Synchronous observations were conducted wherein one instrument was situated at Karakyr and the other 170 km north of it, at Tamdybulak. The synchronous measurements of T were conducted round-the-clock every 5–10 minutes. The recording was executed in a continuous series from 13th to 21st May 1976. The major event during that time was the 9- to 10-degree earthquake of 17th May 1976 ($M = 7.3$).

Basing on the graphs of field changes at reference points of Karakyr and Tamdybulak from 15th to 18th May 1976 (Fig.47) one can draw the following conclusions:

1. Slow field changes occurring some days before the quake and disappearing after it were not present. The field change difference was at one level, with the exception of the instance of bay passage,

2. No field changes of any kind occurred at the time of the quake on 17th May 1976.

The absence of the anomalous effect at the time of the violent Gazli earthquake (9–10-degree) on 17th May 1976 was unexpected.

During the period of monitoring of the magnetic field (13th to 22nd May 1976)

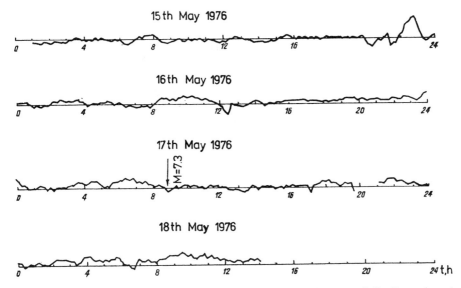

Figure 47. Graphical variation of magnetic field between epicentre of Gazli earthquake and reference of Tamdybulak.

one noted the presence of bay-type variations with characteristic times of one to three hours. The bays at Tamdybulak and the magnetic observatories of Yangibazar, Vannovskaya and Novokazalinsk were identical, and strongly distorted by 85–90% each time at the epicentre (Fig.48).

In the practice of studies on geomagnetic field variations one often encounters similar phenomena (Rikitake 1966a, Stacey & Westcott 1965, Kuznetsova 1969, Abdullabekov & Maksudov 1975). They are usually linked to the anomalous electrical conductivity of rock. Similar field changes can also be associated with processes at the earthquake focus.

From mere observations of 1976 it is difficult to draw unique conclusions on the nature of the anomalous field distortions at the passage time of a bay disturbance. Repeated magnetometric observations by the earlier schedule and method were therefore initiated at the epicentre in the spring of 1977.

During the period of observations (from 16th to 19th April 1977) we recorded different types of geomagnetic field variations, including the amplitude and characteristic times close to anomalous variations in 1976 (Fig.49).

Table 10 contains information on distorted bay-type variations at the epicentre of the Gazli earthquakes of 1976 and 1977. Simple comparison shows that the bay distortion was clearly anomalous in 1976, which seems to be associated with the seismic activity. However, this must not be a final conclusion. Similar phenomena will be further analysed in details.

Investigations of the geomagnetic field variation were also conducted after the

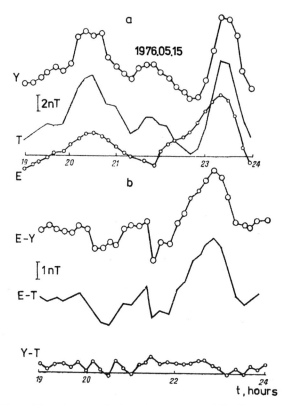

Figure 48. Distortion of bay-type variation at epicentre of Gazli earthquake on 17th May 1976: (a) passsage of bay-type variation at Yangibazar (Y), Tamdybulak (T), and Karakur (epicentre); (b) field difference between stations.

Table 10. Parameters of bay-type variations recorded at the epicentre of the Gazli earthquakes of 1976 and 1977.

Date and time of bay passage	Duration of effect	Amplitude nT	Magnitude of deviation from normal course	
			nT	%
1976				
13 May; 14:20–16:20	2 h 00 min	8–10		
15 May; 19:50–21:10	1 h 20 min	5–6	4	73
15 May; 22:40–24:00	1 h 20 min	11–12	10	87
17 May; 19:20–22:20	1 h 00 min	3–4	4–5	100
19 May; 21:20–22:30	1 h 10 min	14	12	86
1977				
16 Apr; 20:40–23:00	2 h 20 min	22	5	23
16 Apr; 21:40–24:00	2 h 20 min	19	5	26
17 Apr; 00:40–01:50	1 h 10 min	24–29	6	22.5
17 Apr; 22:30–24:00	1 h 30 min	9–10	2–3	25

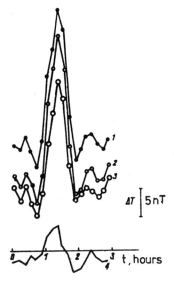

Figure 49. Passage of bay-type variation at epicentre of Gazli earthquake in 1976 (1) and at stations Tamdybulak (2) and Yangibazar (3) on 17th April 1977, and the field difference between the stations Tamdybulak-epicentre (4).

earthquake of 19th March 1984 (Abdullabekov et al. 1986). In 1984 the studies were carried out by the method of stationary all-day measurements at stations Tsvetushchaya, Ozero and Dzhangeldy, which were only slightly aside from the epicentre, and by the method of repeated route surveying. At stationary stations one identified anomalous field changes with an intensity of 2.3 nT associated with individual aftershocks. The measurements on the routes were conducted at a frequency of 1–2 weeks up to several months. In the part of the route cutting across the fault zone the anomalous changes had an intensity of 8–10 nT.

At the epicentre of the Gazli earthquake of 1976 the measurements were conducted during a short period of time, so the results obtained are not definite. The nature of the bay-type variations of the magnetic field before seismic events was later investigated for a longer span of time. Bay-type disturbances manifested in the absolute value of the total vector of the geomagnetic field density (T) with a period from 0.5 to 2–3 hours were used, and the ratio of the amplitudes of the bay-type variations in the investigated area to the reference station was computed (Fig.50).

It is seen from the drawing that the ratio of the amplitudes Chimion vs. Yangibazar (curve 1) falls dramatically from the beginning of February and reaches its minimum at mid-April — the decrease continued for roughly 2.5 months. The value increases afterwards and the original background level is reached in one month. It is also visible that the earthquake took place after the sign had

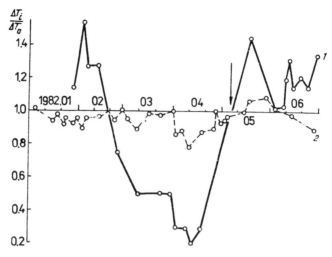

Figure 50. Anomalous changes in the ratio of amplitudes of bay-type variations at stations Chimion (1) and Madaniyat (2).

changed, in the middle section of the fast growth of the amplitude ratio. In the subsequent period, any strong aftershocks of this and other earthquakes in this region were not detected. This is confirmed by the course of the curve. The anomalous change began almost three months before the earthquake and was associated with the distortion of the amplitude of the bay-type disturbances in the region of incipient earthquake. The change was 70–80% of the field disturbance at Yangibazar Observatory. The amplitude ratio Madaniyat vs. Yangibazar, which is given for comparison, is almost at one level, although a slight reduction is also observed here in coincidence with the minimum value of curve 1. It should be noted that no violent or notable earthquake occurred in the Madaniyat region during the observations.

The above material shows that anomalous changes in bay-type variations at epicentres of violent earthquakes are not accidental.

7.2 EPICENTRE OF NAZARBEK EARTHQUAKE OF 1980

On 11th December 1980 an earthquake with $M = 5.5$ and $H = 10$–12 km occurred 195 km south-west of the city of Tashkent, at Nazarbek settlement. Systematic magnetic measurements of the absolute total vector T were initiated on the next day. They employed a self-contained field proton setup; the time step of measurements was 20 minutes, with the accuracy of single readings of 0.4 nT and self-sufficiency of 3 months (Berdaliyev et al. 1980). The stations were deployed at Lenin Yuly (LY) and Nazarbek (NZ), which were located in the epicentral zone (Fig.51).

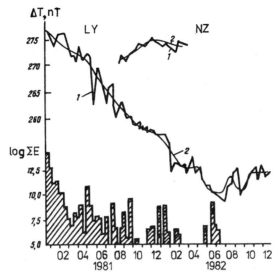

Figure 51. Anomalous changes in magnetic field at epicentre of Nazarbek earthquake of 1980 associated with falling aftershock activity. (1) 10-day values; (2) monthly values. The scheme of magnetometric observations is depicted at top: (1) epicentral stations of Lenin Yuly (LY) and Nazarbek (NZ); (2) = epicentre; (Y) Yangibazar; (BSh) Baka Shel; (NN) Nanay.

Increments of the field differences ΔT obtained from synchronous measurements at a given point and remote 'normal' stations, at Baka-Shel and Nanay (BSh and NN), with a similar apparatus, have yielded the useful signal. The total error of effect discrimination was below 1 nT.

The qualitative comparison of ΔT and $\log E$ shows that slow decrease in the geomagnetic field occurred at Lenin Yuly synchronously with weekening of seismic conditions at the focus. The rate of field reduction was about 2 nT/month, and the field decreased by 27 nT during the time interval. One notes a relationship between the released seismic energy and the local change in the geomagnetic field. At the same time the field change at Nazarbek, 5 km south-west of Lenin Yuly, had the oppposite sign (Fig.51). This bears wittness of diversification of the phenomenon over the area — in fairly good agreement with the model of the generation of crustal earthquake (Dobrovolskiy 1984, and others).

The amplitude of the geomagnetic field changes in the epicentral zone of Nazarbek earthquake has been supported by theoretical estimates found for the conditions of the area (Abdullabekov & Golovkov 1970).

It is interesting to note that the field fall at Lenin Yuly is not steady in time. The period of the field fall can be splitted up into three stages (Fig.51). At each stage.the field change was exponential, with the characteristic time about months.

However, in April 1981 and January 1982 the monotonic course of the curve was disturbed. It should be emphasized that the release of energy along the above segments is accompanied by outbursts of aftershock activity.

7.3 EPICENTRES OF ALAY, TAVAKSAY, CHIMION, PAP AND OTHER EARTHQUAKES

The first epicentral observations after the violent earthquake were conducted in Alay Valley. During several days, from 18th to 23rd August 1974, stationary and repeated repeated route observations were carried out in the epicentral zone. The earthquake epicentre was in inaccessible rocks of the Alay Ridge, so the observations were conducted some 50–60 km away. Both stationary and repeated route measurements were incapable to identify the anomalous changes, and perceptible aftershocks were not detected, either.

Soon after the Tavaksay earthquake of 6th December 1977 ($M = 5.5$), round-the-clock stationary measurements were arranged every 10 minutes at Yangibazar Observatory and the epicenter; and synchronous measurements of T were conducted with the aid of TMP-proton magnetometers. During two weeks of round-the-clock measurements the difference of magnetic field at the epicentre versus Yangibazar Observatory was constant (Fig.52). A certain increase in the difference was observed for a few hours on 16th December 1977, when it rained and the higher difference seemed to be associated with the showers. With exclusion of this change, the magnitude of the field at the epicentre was constant. During the same period, the aftershocks of the Tavaksay earthquake were also imperceptible.

In the case of Chimion (6th May 1982) and Pap (18th February 1984) earthquakes, the stationary measurements were initiated long before the shocks and were conducted after the quakes. In both cases the anomalous changes were bay-type, and a return to the background level occurred after the earthquakes. During the period of earthquake generation one observed an anomalous field fall, while the return to the original level was observed after the quakes in the course of aftershock activities. The fall seems to be associated with build-up of elastic stresses before the earthquakes, while the return to the original level is due to the dissipation of the accumulated energy, i.e. to the aftershock activities. The break in the aftershock activity coincides with the moment when the field magnitude reaches its original level.

Hence the variations of the magnetic field in epicentral areas of the past earthquakes can bear information about subsequent processes at the focuses. The attenuation of the aftershock activity is accompanied by changes in the magnetic field, that is relaxation of the accumulated elastic stresses brings about the return of anomalous magnetic effects to the original level. If aftershock activities do not exist, the field magnitude remains unchanged (for instance, in the case of the Tavaksay earthquake).

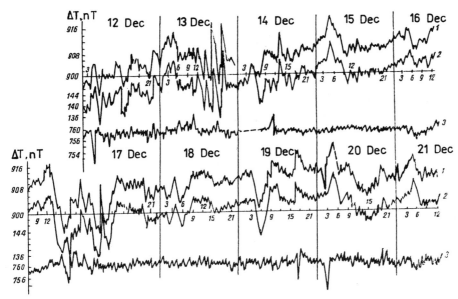

Figure 52.   Variation of magnetic field at epicentre of Tavaksay earthquake on 6th December 1977 (1), and at Yangibazar Observatory (2), together with their difference (3).

7.4 ANALYSIS OF RESULTS

The results of magnetic field measurements in epicentral zones of violent earthquakes given in this chapter are clearly insufficient for sound statistical conclusions.

The measurements were practically conducted in 5–6 epicentral zones only, and the course of magnetic field and aftershock activity changes was different in each zone. At the same time the results were new and unexpected in a number of cases.

The major peculiarity of the Gazli earthquake of 1976 was the absence of the magnetic effect, although the liberated energy was by many orders of magnitude higher than the magnetic field counterpart. Synchronous instantaneous magnetic field changes at the very time of earthquake or a strong aftershock were not observed at the earthquake focus in any single case of continuous record. Even in the epicentral zone of the Nazarbek earthquake, where the long-term magnetic effect was about 40 nT in total — that is by two orders of magnitude higher than the certified accuracy of measurements, the jump in the magnetic field was not present at times of single strong aftershocks having a magnitude of 4.5–5.5. This finding is obviously of fundamental importance in elucidation of the processes at earthquake focuses. We will comment on the validity of this finding in view of our data.

1. Continuous recording of magnetic field was conducted in the epicentral

zones of the Gazli, Nazarbek, Pap, Chimion and other earthquakes when the major shocks or strong aftershocks occurred in them (1976 Gazli and 1984 Pap earthquakes). Sharp changes in the magnetic field of the epicentral area were not recorded at the time of quakes in any single case.

2. Results of the preceding chapter do not contradict the statement on the absence of dramatic field changes. Although in most cases the field was recorded at large distances from the focus, the results confirm indirectly the conclusion drawn as the bay-type changes ended practically before the shock (which enabled us to employ this phenomenon in the prediction of earthquake time). Hence, at least for the crustal earthquakes of Middle Asia, the conclusion on the absence of dramatic changes in the magnetic field in the epicentral zone seems well documented. Results of direct observations at the epicentre of the 10-degree Gazli earthquake are of particular documentary value.

Very interesting, although unfortunately exceptional, are the magnetic field changes at the epicentre of the Nazarbek earthquake. To the best of our knowledge, this is the only case in the world geophysics when the entire aftershock period of a violent earthquake, lasting almost two years, was studied in detail. All data are unique in this case — the total amplitude of the field change was almost 40 nT; it also varied at a distance of 5 km, although in the opposite phase; three cycles of field attenuation due to repeated aftershock activities were clearly seen. A retrospective analysis of other data makes it possible to assume that the phenomenon is rather confirmed than refutable. Indeed, the results given in Chapter 5 for the magnetic field along the route passing 5–15 km from the epicentre of the Abaybazar earthquake provide evidence. The measurements conducted 20–25 days after the shock have shown that the field change accumulated over three years (up to 23 nT) decreased thereafter more than two times, although the form of the anomaly along the route was yet clearly visible (Figures 23 and 24). However, the measurements carried out one year after have proved that the residual anomaly disappeared in practice. Accordingly, if one had a continuous series of measurements at points of that route one could detect the effect similar to the Nazarbek one. Obviously, the effect is confined in space — which follows from both the field change differences at points Lenin Yuly and Nazarbek and from the spatial dimensions of the magnetic effect of the Abaybazar earthquake. This means that it was only a favourable coincidence that enabled us to identify the effect in its whole realm.

The most specific conclusion has already been done in Section 5.2 — the magnetic field does not change monotonically. There exist at least three cycles of the monotonic behaviour, and within each of them the field decreases exponentially in the first approximation, with the characteristic time of the order of 1.5–2.0 months. Afterwords the growth of the aftershock activity interrupts the cycle, and a new one begins. From this it can be concluded that the processes at the focus of the present earthquake behave in a complex manner whereupon at least two characteristic times can be selected — of the order of months (cycles of the magnetic field and the aftershock activity) and 1–1.5 years (total time of the damping of the aftershock activity and the stabilization of the magnetic field).

From the measurements of the magnetic field in the epicentral zone it follows that changes in the bays of the magnetospheric origin with characteristic times of 0.5–3 hours occur up to 70–80% of the seismic activity time.

Another result could be taken with regard to the effects of strong aftershocks of a major earthquake. The epicentres of the Nazarbek (1980) and Gazli (1984) earthquakes have shown anomalous changes in magnetic field, preceding strong aftershocks and phases of aftershock activities. One also observed events typical for periods prior to violent earthquakes. Hence we can conclude that strong aftershocks in a regional magnetic field behave as individual independent earthquakes. This conclusion could seem trivial if not the fact that aftershocks often follow one another much faster than the bay-type disturbances. It is possible that the aftershocks following in the wake of the magnetic bays are a separate class of events. It is not excluded, either, that they themselves are responsible for the type of field fall, as exposed in the epicentral zone of the Nazarbek earthquake, although this particular earthquake has not displayed the bay-type forerunners.

CHAPTER 8

Variations of the Earth's pulsed electromagnetic field

8.1 NATURAL ELECTROMAGNETIC PHENOMENA

A lot of interest has been concentrated on electromagnetic forerunners of earthquakes. They include luminous phenomena, disturbances in voltage of the atmospheric electricity and in geomagnetic field, terrestrial currents and pulses of electromagnetic field, and unusual behaviour of animals and fish, etc.

Extensive information has been provided for the luminous phenomena accompanying earthquakes (Parkhomenko & Martyshev 1975, Rikitake 1979, Ulomov 1971b, 1982, Tserfas 1981, Pribitkov 1973, Sobolev & Demin 1980, Khusamiddinov 1983a, b, and others). In Middle Asia the luminous phenomena were observed during Ashkhabad (1948), Brichmullin (1959), Tashkent (1966), Isfara-Batken (1977) and other earthquakes. In some studies (Parkhomenko & Martyshev 1975, Ulomov 1982), the information on the luminous phenomena is presented in a systematic manner. Mikhalkov & Chernyavskiy (1930) recorded disturbances in the atmospheric electrical field three to four hours before the Kurshkhab earthquakes of 1924 ($M = 6.4$ and $M = 6.2$); they used a noninertial water-jet electrograph apparatus. The amplitude of the pulsations exceeded the 400-W limit of the measuring facility. The period of pulsations was counted in fractions of a second.

Disturbances in the atmospheric electrical field were also observed before the Chatkal (1946), Khait (1949), Tashkent (1966) and other earthquakes in the Soviet Union, and during the Matsushira earthquake swarm (1965–1967) in Japan (Chernyavskiy 1955, Bonchkovskiy 1954, Tserfas 1951, Kondo 1973, and others).

The anomalous variations of the electro-terrestrial field associated with earthquakes have been noted in seismoactive regions of the USSR (Middle Asia, the Caucasus, the Far East), Japan, the USA and China (Tikhonov et al. 1954, Yanagihara & Iosimatsu 1968, Sobolev 1975, Sobolev & Morozov 1970, Sobolev & Demin 1980, Rikitake et al. 1966b, c, Davis 1975, and others).

The relation of terrestrial currents and earthquakes was investigated most fully by Sobolev (1975) in Kamchatka. He identified numerous precursory effects of earthquakes.

Another type of phenomena incorporates the generation of strong electric cur-

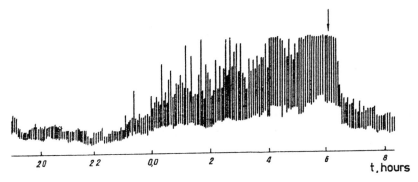

Figure 53. Record of EMR disturbances during 5-degree aftershock of Gazli earthquake of 17th May 1976.

rents during periods of seismic activity. The period of the aftershock activity of the Tashkent earthquake in 1966 favoured the generation of discharges at extremities of cables submerged in 500-m wells. The magnitude of the electric voltage between the deep cable and the surface ground was assessed by Ulomov (1971, 1982) at a level of 5–10 kV.

Sonic phenomena and insulation damages were observed in underground cables laid in granite. As a result of the strong discharge, the ends of the cables were melted (Müller 1930, Khusamiddinov 1983a). Of particular interest in the investigation of processes at the focus of an incipient earthquake is the electromagnetic radiation (EMR) in the band of long and very long radio waves. The studies were first initiated at the Tashkent geodynamical test area. In the years 1972–1974 associates of the Institute for Seismology of the Uzbek Academy of Sciences and the Tomsk Technical University conducted episodic observations on the variations of the pulsed electromagnetic field of the Earth in a borehole of Charvak. It has been evidenced that the Earth's crust emits electromagnetic pulses and that the strength of the radiation decreases sharply before seismic events (Vorob"ev et al. 1976, Mavlanov & Ulomov 1976).

Figure 53 shows an example of record for the so-called pulsed electromagnetic radiation in the band of 12.5 Hz before an aftershock of the Gazli earthquake of 17th May 1976.

The research on the variation of natural pulsed electromagnetic field of the Earth has recently expanded (Anonymous 1982, Anonymous 1983, Vorob"ev 1970, 1976, 1979a, Gokhberg et al. 1980, Sadovskiy et al. 1979b, Mavlanov et al. 1979b, and others). The studies have been conducted in the test areas of Middle Asia, the Caucasus, Kazakhstan, the Far East, western Ukraina, etc and also in the USA, Japan, North Korea and other countries.

In Kamchatka, a few days before an earthquake with $K = 10–11$, anomalous effects in the natural pulsed electromagnetic field were observed within the radius of 200 km in the band of 10–100 kHz (Butakov et al. 1979).

A few hours before the Carpathian earthquake of 4th March 1977, Sadovskiy et al. (1979b, 1980) detected an increase in the intensity of the natural pulsed electromagnetic radiation. The voltage growth of this radiation in superlong radio-waves and medium waves before the Iran earthquake was recorded by Gokhberg et al. (1979), at a distance of 200 km from the epicentre in Dagestan.

The electromagnetic radiation effects identified by various researchers are in two ranges of the electromagnetic spectrum. The first of them is in the very long and long band (wavelength from hundreds to first thousands of metres and frequency from thousands to a million Hz), while the second is the range of visible light.

8.2 TECHNIQUES OF MEASUREMENT OF ELECTROMAGNETIC RADIATION CHANGES IN UZBEKISTAN

Investigations of the variation of the electromagnetic radiation (EMR) associated with earthquakes were conducted in territories of the Tashkent, Fergana and Kyzyl-Kum geodynamical test areas. The measurements were carried out at stationary stations of Yangibazar, Charvak and Andizhan, and in the epicentral zones of the Gazli (1976 and 1984), Nazarbek (1980), Chimion (1982), Alay (1974) and other earthquakes.

A special apparatus designed on the basis of standard industrial units was used in the measurements of the variations of the intensity of pulsed electromagnetic signals (Khusamiddinov 1983b, Abdullabekov 1982). Its sensitivity was below 10 μV, and the range of working frequencies extended from 12 to 1500 kHz. The voltage envelope and the number of EMR pulses were recorded automatically, while counting of the number of pulses was facilitated by the use of a PP-9 counter and BZ-15M printers. The facility was controlled by a contact clock of KPCh-2 type. A high sensitivity superheterodyne radio receivers of R-672 or R-250 type were used for radio reception.

The design of the apparatus made it possible to record independently parameters of EMC, i.e. the number of pulses (N_{EMR}) and the total energy emitted in the range of measurement (E_{EMR}).

The variation of the electromagnetic field measured on the Earth's surface consists of natural and artificial components, the latter being of different origin. The sources of the artificial variations of EMR are usually caused by various industrial factors, industrial noises and radio transmitters, while the natural sources also include atmoshperic lightning charges. In order to identify the anomalous EMR changes associated with processes in the Earth's crust one must include all types of the artificial and natural effects.

Twenty to sixty thunderstorm days are encountered yearly on territories of the

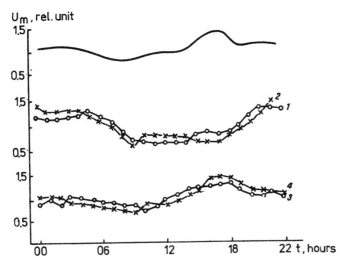

Figure 54. Mean daily course of EMR at stations Yangibazar (1, 3), Charvak (2) and Andizhan (4) in the years 1974–1979; (a) during year; (b) summer; (c) autumn.

Tashkent and Fergana geodynamical test areas. The storm activity is of daily and seasonal type. The maximum of the storm activity during a day occurs at noon, while the yearly maximum appears in April-June.

The identity of the daily course noted at remote stations (Charvak and Andizhan, Figure 54) testifies to the regional nature of changes in the EMR intensity. The daily course of EMR in winter displays one maximum (in the night) and one minimum (in the morning) while those in the spring and summer have two maxima (afternoon and night).

Our investigations have shown that the atmospheric precipitation, sand and dust storms, solar flares, magnetic storms, ionospheric interference and other sources do not distort the magnitude of EMR variations (Mavlanov et al. 1979b, Abdullabekov et al. 1980a, 1982, Khusamiddinov & Abdullabekov 1983a, and others).

The analysis has displayed that the primary sources of EMR of irregular nature consist of close thunderstorms, mechano-electrical processes at focuses of future earthquakes and different artificial noises. Discrimination of the changes associated with earthquakes was attempted in each particular case by elimination of the other types of variation.

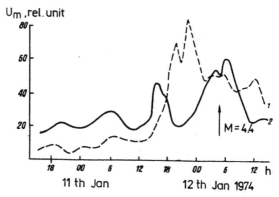

Figure 55. Anomalous changes in normalized EMR field at stations of Charvak (1) and Yangibazar (2) during the Pskem earthquake of 12th January 1974.

8.3 RELATION OF ANOMALOUS EMR CHANGES TO EARTHQUAKES

In order to discriminate the anomalous EMR effects of seismic origin one must include the periodic field components. Basing on the present knowledge of the phenomena, the most intensive anomalous radiation of electromagnetic field pulses occurs immediately before the shock, at the instant of the generation of major fissures. Studies on the short-term field effects in the structure of EMR variation are therefore of utmost importance. In discrimination of short-term earthquake forerunners one can neglect the seasonal changes of EMR. In filtration of the regular daily components we found deviations of mean hourly EMR values from mean monthly course of the daily values. The filtered series were compared with the seismic events.

Consider the most characteristic effects before earthquakes (Fig.55). The earthquake epicentre is situated 40 km from Charvak station and 90 km from Yangibazar station. The field density at the close station of Charvak increased gradually during 10–11 hours, reached the maximum six hours before the quake, and then fell. The effect at Yangibazar was more complex, with two maxima of which the first, about 6 hours long, was observed some 16–10 hours before the shock, after which the field returned to the ground level, and the second anomalous growth began 5 hours before the quake. The anomalous field value was preserved during 6–8 hours and after the shock.

Figure 56 illustrates the variations of the mean daily changes in EMR at Yangibazar from July to August 1974 (Mavlanov et al. 1979b, Abdullabekov et al. 1980). The arrow indicates the time, class (K) and distance (R) from the epicentre of the Alay earthquake of 11th August 1974. Anomalous EMR changes occur on the background of normal variations before the earthquake. The effect

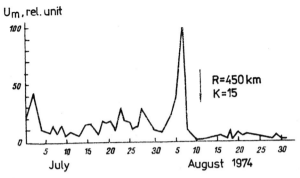

Figure 56. Anomalous variation of mean daily EMR changes at Yangibazar before Alay earthquake of 11th August 1974.

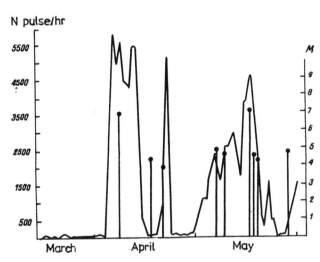

Figure 57. Anomalous changes in mean daily values of EMR at Yangibazar during the Gazli earthquakes of 8th April and 13th May 1976.

occurred 5 to 6 days prior to the earthquake.

At Yangibazar the field intensity did not exceed 10 dB before 3rd April 1976 (Fig.57). An increase in the field intensity was later observed from 5th to 14th April 1976. The first Gazli earthquake with $M = 7.0$ occurred during that time (8th April), 520 km from Yangibazar, and was accompanied by numerous aftershocks (the most perceptible of which are shown in the drawing). The short

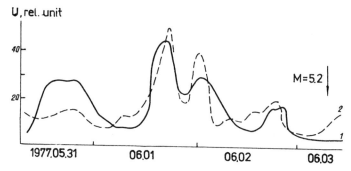

Figure 58. Anomalous EMR effect before Khaidarkan earthquake on 3rd June 1977 at stations of Yangibazar (1) and Andizhan (2).

growth was observed from 20th to 25th April. Two perceptible shocks with $M = 4.5$–5 appeared during that time. Further anomalous field growth was observed from the beginning to 25th May. The second destructive Gazli earthquake with $M = 7.3$ and some strong repeated shocks with $M = 4$–5 occurred during that time (on 17th May 1976). It is practically impossible to identify the disturbances preceding individual events if one uses the mean daily data. The periods of seismic calm were accompanied by falling intensity of EMR.

Aside from the above cases, anomalous behaviour of EMR was also detected during the Isfara-Batken (1977), Kyzyl-Kum (1978), Tyup (1978) and other violent and repeating moderate earthquakes.

As an example one can discuss the case of the anomalous effects (Fig.58) recorded during the Khaydarkan earthquake on 3rd June 1977 ($M = 5.2$) at stations of Andizhan and Yangibazar (Mavlanov et al. 1979b, Abdullabekov et al. 1980a, b, 1982, Khusamiddinov & Abdullabekov 1983b).

The type and time of the anomalous EMR changes are identical at both stations. The amplitude of the effect is slightly higher at the closer station of Andizhan (120 km), compared with the remote station of Yangibazar (200 km). The intensity of the anomaly during nighttime is much higher than during the day.

Hence the intensity of EMR increases anomalously before earthquakes during times counted in tens of hours and days. Earthquakes occur when the anomalous changes attenuate.

The data on particular behaviour of EMR variations during the aftershock activities is also available for the Alay (11th August 1974), Gazli (8th April and 17th May 1976), Isfara-Batken (31st January 1977), Tavaksay (6th December 1977), Nazarbek (10th December 1980), Chimion (6th May 1982), and other earthquakes. The type of anomalous changes before aftershocks is somehow different from the course of changes before the primary shocks. For instance, the

Table 11. Anomalous EMR effects before earthquakes.

Date	ϕ	λ	H, km	M	R, km	T, day	$\frac{A_i}{A_o}$	Station
12 Jan 74	41.6	70.7	5	4.4	90	1	2.0	Yan
					40	1.5	4	
22 Jan 74	40.2	71.1	25	4.7	120	2.5	2.5	Yan
					160	2.5	1.5	Cha
11 Aug 74	39.3	73.7	-	6.7	415	4	3.5	Yan
27 Aug 74	39.3	73.7	-	6.1	415	2	2	Yan
21 Mar 76	40.8	69.7	25	4.4	40	3.5	3	Yan
8 Apr 76	40.4	63.5	20	7.0	525	3.0	2.0	Yan
					750	3.0	2.5	And
10 May 76	40.4	63.4	25	7.3	530	4.5	3.5	
					760	4.0	2.5	And
31 Jan 77	40.0	70.5	25	5.6	150	5.5	3.5	And
3 Jun 77	40.2	71.7	-	5.2	200	1.5	2.5	And
					120	1.5	3.5	And
6 Dec 77	41.7	69.7	15-20	5.3	50	3.5	2.5	Yan
4 Jun 78	40.58	64.0	-	6.2	515	2.5	2.0	Yan
1 Nov 78	39.50	72.60	-	6.4	315	2.0	2.5	Yan

(Yan) Yangibazar; (Cha) Charvak; (And) Andizhan

anomalous effects before the aftershocks of the Gazli earthquakes appeared some hours before the shocks, and had an insignificant intensity and a much shorter duration. The shocks occurred when the field anomaly decreased.

The anomalous changes in EMR and magnetic field before aftershocks of the Nazarbek earthquake were pronounced 1–3 days before and were longer.

Table 11 contains a catalogue of anomalous effects before earthquakes (Abdullabekov et al. 1980, 1982, Khusamiddinov & Abdullabekov 1983, and others). The pronouncement of the effect of crustal earthquakes is clearer than for sub-crustal ones. The anomalous effects are detected at distances ranging from tens to 1200 km. The characteristic time of the effect varies from several hours to 5–6 days. The ratio of the anomalous effect to its background changes from 2 to 4.

The investigations have shown that the dependence of the intensity and time of anomalous EMR on the epicentral distance and energy class does not exist (Abdullabekov et al. 1980, 1982, Khusamiddinov & Abdullabekov 1983b).

8.4 POSSIBLE NATURE OF ANOMALOUS EMR EFFECTS BEFORE EARTHQUAKES

There is no unique opinion on the nature of electric effects. A hypothesis on the generation of strong electric fields in the Earth's crust has been put forth by Vorob"ev (1970). In his opinion, the energy released upon discharges in rock can reach 1.44×10^9 J. The discharge occurs during 10^{-7} s, so the instantaneous power should reach 10^{20} W. It is difficult to agree with Vorob"ev (1970) on the possible initiation of earthquakes by electric discharges, although the hypothesis on the generation of strong electric currents in dielectric rock deserves attention.

Gokhberg et al. (1980) describe the sources of electromagnetic earthquake effects as mechano-electrical (piezoelectric effect, Stepanov's effect, processes of electric charging by friction, failure or destruction of double electric layers) which occur at earthquake focuses, and electrokinetic phenomena, or electric eddy fields. All mechano-electrical processes are sources of EMR due to relaxation of the separated discharges. Piezoelectricity is understood as the phenomenon of electric charging of anisotropic dielectrics due to a mechanical stress applied to them (Sobolev & Demin 1980). In 1953 Volorovich & Parkhomenko (1954, 1955) recorded experimentally and explored theoretically the piezoelectric effect in samples of granite, gneiss and vein quartz.

The phenomena of electric charging, luminous effects and electromagnetic radiation under laboratory conditions were investigated upon destruction and friction in rock. Numerous experiments have shown that the generation of new surfaces upon failure and friction of materials is accompanied by light flares and electromagnetic radiation in a wide band of frequencies (Gokhberg et al. 1982).

Generation of electric charges in many materials (Nacl, KCl, LiF, AgCl, MgO, CaF_2, feldspar and others) which do not possess piezoelectric properties has been discovered by A.V.Stepanov. In this phenomenon, an electric charge appears at places with non-uniform deformation, while unloading brings about relaxation of the discharge during fractions of seconds (Sobolev et al. 1982). Stepanov's effect was investigated in detail on feldspar samples, this rock being widespread (Shevtsov et al. 1975). Under uni-axial strain at a rate of 0.2 mm/s, a charge of 10^{-5}–10^{-9} C/m^2 appears on the surface of the sample. The current before the damage reached 10^{-5}–10^{-8} A/m^2.

Sobolev & Demin (1980) have provided a physical background for the behaviour of forerunners by reference to the fissuring hypothesis on generation of earthquakes.

Khusamiddinov (1981) believes that generation of EMR takes place in the atmosphere above an earthquake focus or in a subsurface layer of the Earth's crust. EMR signals propagate primarily in the atmosphere. Their generation at a considerable depth and channel type propagation via natural waveguides in the Earth's crust are not excluded.

Field studies on the variation of EMR electromagnetic radiation, atmospheric electrical field, electroterrestrial field, and luminous and ionospheric effects are described in the monograph on electromagnetic forerunners of earthquakes (Anonymous 1982), together with results of laboratory studies on the mechano-electrical processes in rock. The monograph also contains a description of the electromagnetic radiation based on contemporary concepts on the physics of the earthquake focus, and some other problems.

Analysis of the results obtained in Uzbek test areas and some other seismoactive regions (Korneychikov 1985, Kurskeyev & Korneychikov 1984) and under laboratory conditions, with inclusion of earthquake generation models, makes it possible to elucidate the anomalous variation of pulsed electromagnetic fields in the following manner, basing on the present knowledge. In the most brittle part of the Earth's crust, when tectonic stresses build up, different discontinuities of rock appear and electric fields of high voltage are generated. These fields inside the Earth's crust bring about additional currents and cause changes in the usual picture of the electroseismic and magnetic fields.

CHAPTER 9

Comprehensive analysis of results of electromagnetic measurements in Uzbekistan

In our research on anomalous changes in the magnetic and pulsed electromagnetic fields associated with various processes in the Earth's crust we have been guided by two general assumptions:

1. Ferromagnetic rock changes its magnetism upon variation of elastic stresses.

2. The generation and growth of earthquake depend considerably on the build-up of elastic stresses at the focus.

Hence we have relied on the classical physical piezomagnetic effect and, in terms of geophysics, on the mechanism of elastic transfer and growth of earthquake.

Our first results, which proved to be a novelty in the Soviet Union as well, go indeed fully along the lines of the model. This has been confirmed for the Abaybazar earthquake, where the growth of the field in the focus zone and its reduction at the periphery (Figs 29, 30) have been established by repeated measurements in a network of stationary points. The process took more than one year, and the amplitude at the maximum was above 20 nT. The anomalous field disappeared partly after the earthquake — in good agreement with results of laboratory experiments, which pointed to reversible and irreversible changes in rock magnetism (Abdullabekov & Golovkov 1974, Abdullabekov et al. 1972, Abdullabekov & Maksudov 1975). Subsequently similar anomalous changes were also identified before the other earthquakes of Khalkabad (1972), Tavaksay (1977), Nazarbek (1980), Gazli (1976), Alay (1978), Isfara-Batken (1970), and others (cf. Chapter 5, Table 8). At that stage of research we encountered some unexpected results as well. For instance, we have observed anomalous field changes associated with the Prealpine fault in central Europe. It was only roughly that we estimated the time and amplitude of the field changes by data of European observatories (Abdullabekov 1972, Abdullabekov & Golovkov 1974). During 15 to 20 years the general field change reached above 100 nT. This phenomenon was later investigated in detail by Maksimchuk (1983) in the USSR and Mund (1980) in the GDR. These results have confirmed our findings and provided a more accurate description of field changes in the space. At the same time Shapiro (1981, 1983, 1986) was first to identify such phenomena in the Ural Mountains (Monchazh anomaly).

Interpretation of the above phenomenon cannot be given along the classical lines

116

of seismomagnetic concepts. From the point of view of the piezomagnetic effect, the amplitude and spatial dimensions were fairly large, while in geophysical sense they were not accompanied by calamitous earthquakes. In order to explain such field changes we have formulated a hypothesis on regional restructuring of the elastic stress field (see Chapter 1). The magnetic aspect has been included in this hypothesis through a slight increase in temperature upon adiabatic compression of rock at the Curie isotherm depth, which brings about transition to the ferromagnetic condition of small rock masses under small vertical temperature gradient; these masses being sufficiently large to explain the field change. We have also accounted for the anomalous increase in the magnetic susceptibility about the Curie point.

The geophysical interpretation of the results obtained was later presented together with Golovkov & Nurmatov. In a series of studies (Abdullabekov et al. 1981, 1982, 1983, and others) we have shown that the conditions of seismic activity have similar characteristic times and spatial dimensions. This is in good agreement with recent results published by Nikolayev (1984) and other authors who employed entirely different methods.

Another incompatibility has been detected upon measurements above the Poltoratsk anticlinal structure into which natural gas was pumped at an excess pressure above 3 MPa (Abdullabekov & Golovkov 1971, Abdullabekov 1972, Abdullabekov & Golovkov 1974, Abdullabekov & Maksudov 1975).

The magnetic field changes were indeed detected although they were unexpectedly high. The sources of the anomalous field were found in layers of week magnetic sedimentary rock. The piezomagnetic effect could not be employed to produce the detected magnetism (Abdullabekov & Golovkov 1974, Abdullabekov & Maksudov 1975). In order to explain the phenomenon we were forced to use a hypothesis on the electrokinetic character of the field change (Abdullabekov & Sultanbekov 1976, 1978).

The essence of the hypothesis consists in the observation that, upon seepage of an electroconducting liquid through a capillary network of loose rock in the Earth's magnetic field, the induction currents are extremely large (see Chapter 2). They are higher by several orders of magnitude than those which would arise in an electroconducting liquid moving in a magnetic field with the same velocity. The nature of the phenomenon is hardly explored, but many laboratory investigations (Anonymous 1982, Sobolev & Demin 1980, and others) have permitted empirical relationships, the application of which to our conditions has provided a satisfactory elucidation of the effect.

Slightly later, when a network of permanent proton magnetometers (MPP, TMP, APM, etc) has been deployed in Uzbekistan we have collected a bulk of evidence for the essential role of the electrokinetic effect in the generation of anomalous geomagnetic changes. This has been most distinct in the light of the observations on bay-type field changes in the Fergana area (Mavlanov et al. 1979, Shapiro & Abdullabekov 1982, and others). The Paleozoic basement lies here under a thick stratum of weakly magnetic and non-magnetic Mesozoic-Cenozoic sedimentary rock up to 10–12 km thick. A number of bay-type field changes have been detected within the test area before strong earthquakes, the amplitude of which reached 23 nT, and the characteristic time of which ranged from several days to several weeks, while the spatial dimensions were tens of kilometres or more (Table 9). This spatial anomalous field changes have obviously be attributed to particular structural units of the valley.

In combination with geophysical tests in the area of studies we have also conducted extensive hydrological and hydrogeochemical measurements, in cooperation with the Institute for Seismology of the Uzbek Academy of Sciences and under a joint programme (Sultankhodzhayev et al. 1980, 1985). The characteristic property of this type of variation consists in the large area of the effect for a rather moderate dependence of the intensity of the anomaly on distance. The non-uniformity of the anomaly in terms of the gasochemical composition and groundwater level is essential, for they are associated with the permeability of rock.

The bay-type anomalous field changes before earthquakes, detected and investigated jointly with EMR anomalies have provided very reliable characteristics for determination of earthquake time (Fig.59). The same data forces us to reject the model of elastic yield at earthquake focus. The magnetic data has been almost unique among other geophysical background to clearly identify the presence of the particular qualitative stage in the generation of earthquake. Seven to ten days before the event one observed a change in rock condition at the focus which was not of elastic nature, as confirmed by the magnetic data. Unfortunately, magnetologists were unable to specify the nature of these changes. Laboratory experiments on the magnetism in the inelastic stage such as avalanche fissuring or anything of this kind, as they appear in the contemporary models of focus, have not been carried out so far.

As mentioned above, the field changes are bay-type, and the earthquake occurs about the time when the field returns to the background level.

Hence one can distinguish two phases in the growth of focus. In the beginning, for a few days and weeks one observes a certain disturbance in the rock condition which is accompanied by magnetic changes. Thereafter the condition also changes and the magnetism returns to the normal level. It is the latter stage which involves a new forerunner — strong pulsed electromagnetic radiation which is recorded about 12 kHz.

The detection and investigation of the electromagnetic pulses immediately before earthquakes, arising in the sign change phase in magnetic field anomalies have made up the general picture of the electromagnetic phenomena associated with dynamic processes in the Earth's crust. This sphere of phenomena cannot be explained in unique physical terms. Any fissuring in dry rock obviously brings about electromagnetic pulses. From laboratory experiments one also knows the piezoelectric effect as well as changes in electrical resistivity due to elastic stresses (Sobolev & Demin 1980, Parkhomenko & Barsukov 1972, and others). It is also

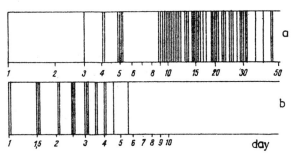

Figure 59. Spectrum of duration of short-term magnetic (a) and pulsed electromagnetic (b) forerunners of earthquakes.

obvious that a part of energy at the focus is converted into heat to warm up the rock as the result of the earthquake (Michelson et al. 1980, Shapiro 1985, and others). Disintegration of rock at fault is accompanied by an increase in electrical conductivity due to the property of electrical conductance of liquid. The latter group of phenomena can be harnessed to elucidate the results of measurements in the zone of Gazli earthquake where particular magnetospheric-ionospheric properties of the bay-type variation appeared during the period of seismic activity. However, in order to explain the pulsed electromagnetic phenomena one must also shed light on their transfer from the focus to the Earth's surface (Gokhberg et al. 1980, Anonymous 1982, 1983, and others). The propagation of high frequency oscillations to the surface indeed reguires a very low specific electrical resistivity of rock above the focus. A number of authors have proposed the waveguide hypothesis, but an ultimate solution has not yet been found.

Finally, our data shows that the instant of earthquake itself is not associated with sharp changes in the geomagnetic field. Even at the epicentre of the 10-point Gazli earthquake of 1976 essential field changes were not discovered at the time of quake.

Hence the above discussion shows that the magnetic model of dynamical phenomena in the Earth's crust of seismic regions can be formulated. Slow but fairly intensive field changes with a characteristic time of 10 years and more have regional character and are associated with tectonic structures of the type of piedmont faults, large anticlinal rises and other mountainous structures with characteristic dimensions ranging from hundreds to one or one and a half thousand kilometres. In the time domain these field changes are correlated with seismic changes of the region, redistribution of earthquakes between latitudinal and meridional systems of faults, etc.

On the background of small changes there arise transitional (mesoscale) anomalous field changes having dimensions up to 20–30 km and more, with a characteristic structure of small isometric bodies magnetized in the direction of the major field of the anomaly. Their form, time and amplitude correspond most fully to the model of elastic stress build-up at the focus of a future earthquake. They are accompanied by regional anomalous changes which, in plan view, represent a combination of positive and negative field effects, the total stretch of which can count in several tens of kilometres. Dimensions of anomalies of one sign determine the dimensions of the tectonic blocks of compression or tension zones.

A short-period bay-type field disturbance arises in a few weeks, or less, and is sometimes detected at a distance of tens or hundreds of kilometres from the focus of the future earthquake. The form of each specific anomaly depends on the inhomogeneity of the structure of the regional crust. In time and space this phenomenon is well correlated with hydrogeochemical changes.

In the descent phase of the bay-type disturbance there arises a new phenomenon — intensive pulsed electromagnetic radiation. At the instant of earthquake EMR is arrested, first and foremost; secondly, it is not displayed in magnetic field changes.

Aftershocks of major earthquakes do not differ from independent earthquakes with regard to a number of symptoms. They can also be preceded by bay-type field disturbances and pulsed electromagnetic radiation. The aftershock activity is however visible in a specific way in the magnetic field of the near field. First, the accumulated excess magnetism of the focus disappears not at the instant of earthquake but during the period of the aftershock activity, and its characteristic time is counted in months. Secondly, this fall can appear on the background of 'breakaway', when the break-down of magnetism begins again after a strong aftershock and an outburst of aftershock activity.

Hence we have investigated four types of anomalous field changes — slow, mesoscale (transitional) and fast, appearing before an earthquake; and the fourth group of aftershock changes. Each type of the anomalous changes has been considered separately, since specific methodology was required for each of them. In principle, all the above types, with the exception of slow changes, reflect different phases of earthquakes, so they can be analysed jointly. It should be recalled that anomalous changes associated with the aftershock activities are forerunners of the focus processes and are inseparable from the transitional and short-term processes. Therefore we shall attempt a comparison of the obtained results with the best known models of earthquake focus formulated in the recent years such as unsteady avalanche fissuring (Myachkin et al. 1974), dilatation-difussional (Sholtz et al. 1973), and qualitative as well as quantitative models postulated by Dobrovolskiy (1984).

The intensities and characteristic times of anomalies can be explained satisfactorily in terms of the avalanche, dilatation-diffusional and Dobrovolskiy's quantitative models. The other properties of the effects such as linear dimensions, spatial distribution and sequence of positive and negative anomalies (mosaic properties) are better described in terms of Dobrovolskiy's model. The same holds true for the physical mechanism of the generation of anomalous changes. The piezomagnetic and piezoelectric anomalies are clearer in the avalanche model while the electrokinetic anomalies are more heuristic in terms of the dilatation-difussion model, and

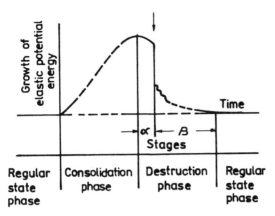

Figure 60. Schematic representation of Dobrovolskiy's consolidation model (1984).

both mechanisms are explained satisfactorily by Dobrovolskiy's model.

The character of the three types of anomalous field changes has also been analysed in terms of the above models. The mesoscale and short-term anomalies are generally compatible with the avalanche model, dilatation-diffusion model and the consolidation model put forth by Dobrovolskiy (Fig.60). The mesoscale anomalies of magnetic field coincide with the consolidation phase while the short-term anomalies of the magnetic and pulsed electromagnetic fields are given for the α-stage. The aftershock anomalies can be attributed to the β-stage of the failure phase within the consolidation model. The forms of the magnetic anomalies are however more sophisticated than those represented by the model. This is essential since the model is somehow idealized. More substantial differences also appear. In the magnetic field the α-stage can have bay-type reversible anomalies while the model produces jumpwise and irreversible changes. The form of the magnetic anomalies due to aftershocks is also more intricate than that simulated by the model. The magnetic field anomaly attenuates as a multi-phase exponential function.

CHAPTER 10

Magnetometric prediction of earthquakes

Magnetometric prediction of earthquakes is a constituent of a general technique aimed at identification of anomalous changes in the Earth's magnetic field associated with earthquakes, and their subsequent interpretation. As known, the emphasis in the investigation on the seismomagnetic effect is placed on the measuring techniques and the separation of anomalous changes, i.e. assessment of instrumentation error, measurement error, and effect discrimination error. In guidelines and methodological recommendations on seismomagnetic measurements one deals with scales of magnetic surveying and requirements on the selection of parameters involved, and optimization problems for a network of routine observations and computations but still some other parameters are given less attention (Golovkov et al. 1977). Hence elaboration of the technique for magnetometric prediction of earthquakes is a part of the methodology of the research on the seismomagnetic effect in geodynamical test areas.

10.1 OPTIMIZATION OF ROUTINE MAGNETOMETRIC NETWORK

A successful prediction of earthquakes by the magnetometric method is based on discrimination of the complete picture of spatial and temporal parameters of anomalous field changes caused by an anticipated earthquake.

Optimization of the network of routine magnetometric observations consists in determination of a minimum number of points or a minimum scale of surveying with which one obtains a complete picture of the anomalous magnetic field changes. The policy of expanding the number of stations and points with the aim of collecting a larger bulk of data is uneconomical and unrealistic from the physical point of view. Anomalous changes due to each earthquake are individual and unique, and the problem must not be treated stage by stage (determination of the effect at individual points first and then detailed description with reference to real dimensions of the detected anomalous changes). The network should be optimal; the optimization can be accomplished in the following ways.

1. Through assessment of the real dimensions of earthquake focuses and display ranges of anomalous changes;

2. Upon thorough consideration of the presence of seismogenic zones and display areas;

3. By establishment of a combined double-scale network (reference and regular);

4. Through evaluation of real dimensions of expected anomalies basing on statistical analysis of data banks for contemporary and recent movements of the Earth's crust, dimensions of blocks, anomalous geophysical fields (magnetic, gravitational, etc), seismoactive zones, etc.

In the selection of a network basing on dimensions of earthquake focuses and display ranges one employs the relationships derived by Dobrovolskiy et al. (1980), Ulomov (1977) and Sadovskiy et al. (1979). Starting from the distributions of elastic stresses in the zones of earthquake generation, Dobrovolskiy et al. (1990) proposed the following formula for the assessment of the display range of earthquake forerunners: $R = e^M$ km. Accordingly, the network should have the dimensions $L < e^M$.

Ulomov (1977) evaluated the dimensions of the display range of earthquake forerunners by reference to multiples of focus dimension. With increasing class (magnitude) of earthquake the focus dimension increases, and so does the display area of anomalous field changes. Analysis has shown that the electromagnetic forerunners are displayed within the range up to 30 focus dimensions (Chapters 5, 6, 8). Therefore one should take $R \leq 30P$.

Sadovskiy et al. (1979) propose that the display zone of anomalous field changes be taken as the area in which the intensity of an anomaly exceeds two standard deviations of the discrimination error of the anomalous effect.

All methods aimed at identification of earthquake forerunners are general and approximate. They hold true upon selection of earthquakes for comparison of anomalous changes in the measured parameters against seismicity. In earthquake prediction, however, for each region or point of observation one encounters characteristic local properties. The presence of a relationship between an anomalous effect and earthquakes can be unique only if real data provides evidence. As an example one can examine the distribution of the display radius of magnetic anomalies versus earthquake magnitude depicted in Figure 61.

It is visible that most magnetic anomalies discriminated in Uzbekistan are contained inside those zones although anomalies outside them are also noticeable. This means that magnetic effects in some cases can be displayed at larger distances than those linked to the effects involved in other methods. In the first approximation, the real display zone of magnetic anomalies can be given by the formula $\log L = 0.46M + 0.08$ (Fig.61). Increasing display range of the electromagnetic effects can be attributed to specific features of identification methods (Chapter 9).

The identification of the display area of electromagnetic forerunners proposed by Dobrovolskiy et al. (1980) is most recommendable in optimization of the forecast network. It is seen from Figure 62 that the real range of anomaly display is undoubtedly within the radius $R = e^M$ km. The selection of the network depends

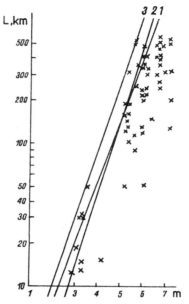

Figure 61. Distribution of magnetic anomaly display radius versus earthquake magnitude: (1) $R = e^M$ km (Dobrovolskiy 1980); (2) $R = 30P$ (Ulomov 1977); (3) $\log R = 0.46M + 0.08$.

on the aim of routine measurements. The most valuable prediction of seismic events begins for $M \geq 5$. Accordingly, the network of stations should be spaced by not more than 150 km one station from another (Fig.62).

The network existing in eastern Uzbekistan (Fergana Valley, Tashkent region) makes it possible to identify earthquake effects for $M \geq 5$, while the stations in southern and western Uzbekistan do not permit discrimination of earthquakes within the entire territory. The existing network allows one to recognize effects 'without target omission' only for earthquakes with $M \geq 6$. Detailed analysis of the seismomagnetic effect requires a denser network. This requirement is complied with in eastern Fergana and at Charvak where earthquake effects for $M \geq 3.5$ can already be distinguished.

In optimization of observations and selection of routine measurement stations one must account for the presence of seismogenic zones. As a rule, strong earthquakes are linked to seismogenic zones. They are elongated narrow segments of the Earth's crust and occupy roughly one-third of the total area of investigations. Routine observations in Uzbekistan are associated with seismogenic zones. This is not very important in identification of the effects associated with strong earthquakes having $M \geq 6$, but plays a substantial role in the case of medium or particularly weak earthquakes. In detailed studies on the seismomagnetic effect, upon arrangement of stations in seismogenic zones, the number of measuring stations can be reduced to 50% or less.

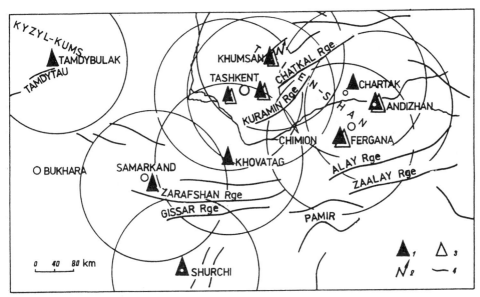

Figure 62. Optimum network of forecast stations in Uzbekistan for earthquakes with $M \geq$ 5: (1) magnetometric stations; (2) dipole electric probing stations; (3) electromagnetic radiation stations; (4) mountain ridges.

The use of a double network permits substantial reduction of the scope of field measurements (Golovkov et al. 1977, Abdullabekov 1972). The reference and accurate regular network points are deployed at the same time. The reference stations are used alone until any anomalous changes arise. If the latter appear the entire network is used and the complete spatial picture of field changes is obtained. The most successful example is provided by the anomalous changes detected before the Khamzaabad earthquake of 1984 (Mumunov et al. 1977).

In optimization of the prediction network it is promising to evaluate the radius of anomalous field display by reference to the real distribution. In this case, however, statistical data must be collected over a few years.

Optimization of measurements and data processing. Optimization of measurements is reduced to the selection of the parameter of measuring network (Δt) which characterizes the repetition of measurements at regular points and the discrete increment in the case of stationary measurements.

In order to obtain the complete temporal picture of field changes one must comply with the condition

$$\Delta t \geq \frac{t}{10} \qquad\qquad (10.1)$$

in which
  t = characteristic time of anomalous effects,
  Δt = frequency of data acquisition.

The quantity t for slow changes is 10–25 years, versus 2–3 months up to 2–3 weeks for the mesoscale (transitional) ones (Chapter 9). The discretization depends on the method of measurements.

In continuous or discrete measurements of the field magnitude under stationary conditions one can use the same method of investigation for the three types of changes. In this case it is however necessary to conform to the condition of optimum data processing.

Fast (short-term) effects are discriminated by superposition of hourly differences between a reference and a measuring station and by construction of time series for the hourly and daily differences. The mesoscale (transitional) effects are found through derivation of mean values for the daily, half-monthly and monthly differences from the aforementioned field-measuring station couples. Slow anomalous field changes are discriminated in the best way if the hourly differences are examined for the network of stations or observatories.

In the study on anomalous changes in the geomagnetic field of crustal nature by repeated route and area measurements it is necessary to pay attention to optimization of measurements. If identical methods are used for mesoscale and slow field changes, the discretization should obviously be indicated as above, that is $\Delta t = \frac{t}{10}$.

One of important problems also involves optimization of measuring accuracy, which is known to depend on the technique of data processing and field changes of different origin, aside from the type of apparatus. At present we are using the Soviet magnetometers MPP, TMP and others having an accuracy of 0.1–0.3 nT, which provide reliable information for the anomalous effects with an intensity ranging from units to tens of nanoteslas. For adequate discrimation of useful signals from noise it is however necessary to distinguish clearly the solar daily and other variations. An important role is also played by the base station. It is known that closer spacing of the reference and a measuring station brings about lower error level due to non-identical source of solar daily variations; magnetic effects are however confined in space, too. Reduction of the spacing between the station couples decreases artificially the intensity of the magnetic effect, and transition takes place from recording of the effect to recording of its gradient. In order to obtain the absolute effect the base station should be outside the zone of field anomaly. Increasing spacing of the station of measurement and the reference (inclusion of variations) brings about higher discrimination error for the anomalous changes.

The experience of many years of investigations shows that the most optimum spacing for Uzbekistan stations is 100–120 km. One must also take into account the geoelectrical conditions of stations. For the above distances they should be identical; otherwise the spacing should be much smaller or the solar daily variation of the magnetic field must be taken into consideration.

10.2 MAGNETOMETRIC PREDICTION OF EARTHQUAKES

In bibliographical sources, the appearance of precursors is often identified with the prediction of time of earthquake, and the percentage of predicted earthquakes grows. As a rule, the prediction of the place and strength of earthquake is not available. Such an earthquake forecast can be regarded as the past of research. Indeed, in the sixties and even at the beginning of the seventies one often referred to reliable precursors, that is the presence or absence of effects. In investigations of that period the discrimination of reliable anomalies before a seismic event was considered an earthquake prediction. At present, the magnetometric prediction of earthquake is understood as the forecast of the place, force and time of expected earthquake basing on interpretation of the time series of magnetic field changes collected in a network of forecast stations.

Division of the earthquake prediction in long-term, mesoscale and short-term (operational) components is done in the time domain. The long-term forecast involves seismic events with an accuracy from 5–10 to several tens of years, the mesoscale prediction being undertaken for months and some years, while the short-term forecast covers the periods from hours and days to 2–3 weeks (Sadovskiy & Nersesov 1978).

The magnetometric method of earthquake prediction also includes a number of methodological problems of seismomagnetic observations in seismic regions such as selection of the region, route, and spot of observation, the scale of surveying, apparatus, the method of measurement, inclusion and evaluation of aseismic changes, the methods of processing, etc. Our programme is largely simplified because these problems have already been discussed in methodological quidelines on the organization and implementation of seismomagnetic observations (Golovkov et al. 1977, Abdullabekov 1972, and others). We will therefore dwell on the problem of prediction for the place, intensity and time of seismic events basing on the interpretation of data collected in a network of magnetic forecast stations.

The prediction takes place in a few stages. In the first stage, data is collected in a network of forecast stations in continuous or discrete series of magnetic records, and graphs are constructed for hourly, daily, weekly or ten-day, monthly, etc time series of field differences measured between a forecast station and its reference, along with time series of differences for couples of forecast stations in different combinations. The accuracy of the differences computed is assessed and anomalous changes in the magnetic field are identified. As known, the criterion of anomaly is taken as the magnitude of the effect above the constructed standard deviation of precursor discrimination, $\Delta T_{aH} \geq 3\sigma$.

Hence the first stage can be referred to as the stage of identification of anomalous changes in the geomagnetic field of crustal nature.

The second stage consists in revision of the anomalous field changes. Obviously, there can be many causes of the anomalous field changes: natural and artificial noises, anthropogenic processes, instrumentation error, computational errors, etc.

Therefore the anomalies found must be carefully inspected. It is only after a fixed reliability criterion is established that the next stage begins. In particular cases additional measurements about the station may prove necessary.

The third stage consists in earthquake prediction with inclusion of the three parameters of time, intensity and place.

Prediction of time. The anomalous field changes before a seismic event have a complex form. In the first approximation they are however similar to a convex or concave bay (positive or negative). Analysis of the relationship between anomalous effects and earthquakes executed for above 100 cases shows that the instant of earthquake usually coincides with the phase of field return to the original level after the field anomaly changes its sign. Figure 63 depicts a histogram of the distribution of earthquake instants versus the sign change of field anomaly. It is seen that the earthquake instants for short-term and mesoscale precursors are mostly (in about 90% cases) at the phase of the return of the field anomaly to its original level. This means that the above regularity can be associated with an expected seismic event. In terms of characteristic times of short-term and mesoscale magnetic precursors, with inclusion of the fact that the first phase of precursors is roughly two-thirds of the time of the first phase, the time of mesoscale events can be predicted with an accuracy of 0.5 to 1–1.5 year. For the short-term prediction the accuracy varies from several hours to ten days.

The accuracy of earthquake forecast depends on the intensity of expected earthquake. It is known that the stronger the earthquake the longer the time of its generation and the lower the accuracy of mesoscale prediction. The accuracy of short-term forecast does not depend on the intensity of earthquake as no relationship has been found thus far between the time of short-term effects and the earthquake intensity.

Prediction of earthquake intensity. An intensity of expected seismic events can be predicted on the basis of the duration of magnetic effects and linear dimensions of discriminated anomalous changes. In literature one encounters a lot of references on the relationships between the duration of precursors and the earthquake energy (Sidorin 1979, Myachkin et al. 1975, Rikitake 1979, Sholz et al. 1973, Whitcomb et al. 1973, Tsubokawa 1969, 1973); cf. Figure 37. A similar relationship of this type has also been established for Uzbekistan (Fig.61):

$$\log T = 0.46M + 0.08. \tag{10.2}$$

One should recall that the relationships found are suitable only for the mesoscale precursors. There has been no relationship for the short-term anomalies.

Another important factor useful in the prediction is the linear dimension of the detected anomalous changes. As shown above (see Chapters 5 and 6), the linear dimensions depend on the earthquake intensity. The stronger the seismic event the larger the area of its generation. Accordingly, the dimensions of anomalies

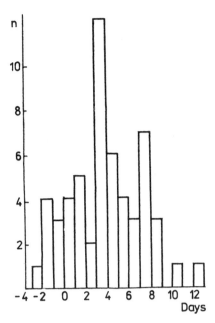

Figure 63. Histogram of earthquake instant versus time of sign change in field anomaly.

carry information on the intensity of an expected earthquake. In the first approx-imation, the magnitude of earthquake can be determined by the formula given in Dobrovolskiy et al. (1980):

$$R = e^M \text{ km.} \tag{10.3}$$

In this case R denotes the display radius of magnetic anomaly and is found directly from an analysis of magnetic data collected in a network of forecast stations.

From the analysis of multiyearly data one has found the following relationship for Uzbekistan (Sec. 10.1):

$$\log L = 0.46M + 0.08. \tag{10.4}$$

The accuracy of earthquake prediction depends on the accuracy of the determi-nation of linear dimensions of anomalous changes and the duration of anomalous field changes.

Analysis of experimental data shows that the earthquake intensity can be predicted with an accuracy up to $M \pm 1$.

The accuracy of the determination of the linear dimensions depends on the number of forecast stations while the accuracy of the determination of time is not associated with that factor.

Prediction of earthquake location. Anomalous variations of electromagnetic effects possess a number of specific properties which make difficult the prediction of a place of future earthquakes. The difficulties involve the absence of any clear relationship between the intensity of effects and the earthquake intensity, the absence of linear increase or decrease versus epicentral distance, the attraction of maximum field changes to indication areas, the existence of a relationship between the intensity of anomalies and geoelectrical conditions and physical properties of rock, etc. Therefore the accuracy of the earthquake place prediction is very low.

The place of a forthcoming earthquake can be predicted by the use of spatial dimensions of anomalous effects. As shown above, the linear dimensions of anomalous changes depend on the earthquake intensity and dimensions of focus zones. Indeed, the stronger the earthquake generated the larger its area of display.

In the practice of earthquake prediction one usually provides contours of station regions with anomalous field changes at the place of expected events as identified with the centre of the contour zone or the entire zone itself. The accuracy of the identification of earthquake place by this method is very low and can reach a few hundred kilometres. The accuracy can be increased if one uses the following simple procedure, which consists of a few stages.

In the first stage the well-known relationship between the time logarithm and earthquake energy is used for each station so that the intensity of the expected earthquake is determined.

In the second stage the relationships $R = e^M$ km (Dobrovolskiy et al. 1980), $\log L = 0.46M + 0.08$ or tabular data from Ulomov (1977) are used to determine the mean radius of display of anomalous field changes.

In the third stage the effect display radius is drawn for each station of anomalous field change. The overlapping zone of the circles around three or more stations becomes the place of expected seismic event in the examined region.

Experience of earthquake prediction in Uzbekistan. Since 1976–1978 the data of combined routine measurements have been used in Uzbekistan to predict earthquakes. In March 1978 a Forecast Committee was established at the Institute for Seismology of the Uzbek Academy of Sciences. The data of daily observations have been processed on operational basis and have been presented at regular weekly meetings of the Forecast Committee. Reports and predictions for the coming week are being prepared on the basis of the results of data processing. In 1978–1980 the first methods for earthquake prediction were elaborated which were later improved and refined.

In 1980–1985 the data of ten stationary magnetic stations were used for earthquake prediction. For operational prediction the magnetic field data measured daily at forecast stations were transmitted by teletype or telephone to the Data Centre, where the differences ΔT for each station versus Yangibazar were computed routinely. The spatial and temporal variations of ΔT were determined, and the expected seismic events were reported on at weekly meetings of the Forecast Committee.

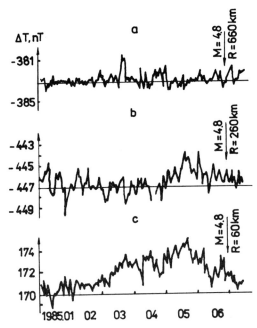

Figure 64. Mesoscale magnetic field anomaly due to Khamzaabad eearthquake of 28th June 1985 at stations Tamdybulak (a), Khavatag (b) and Chimion (c).

From July 1982 to January 1985 the Forecast Committee used the magneto-metric data to provide written evidence of 48 forecasts of expected earthquakes with indication of place, time and intensity (Muminov et al. 1986). The pre-diction of place was regarded as indication of the territory of geodynamical test area, seismic region or individual parts thereof, or areas within a certain radius around any forecast station, in an indicated direction. In 33 cases (about 70%) the prediction was verified with respect to three parameters, in ten cases (about 20%) one of the parameters was not predicted, upon satisfactory prediction of the two remaining parameters, in five cases (about 10%) two or more parameters were not predicted correctly, and in single cases the earthquakes encountered did not display any changes in the magnetic field.

Consider as an example the prediction of the place, intensity and time of the Khamzaabad earthquake of 28th July 1985 with $M = 4.8$ (Muminov et al. 1977) (Fig.64).

From mid-March anomalous field changes appeared at Chimion station. In May 1985 the intensity was 4–5 nT. The anomaly at Khavatag station (about 200 km from Chimion) was discovered much later, and its amplitude was smaller; it was not present at remoter stations. Accordingly, the anomaly was local, and the source was situated about Chimion. In order to predict more accurately the

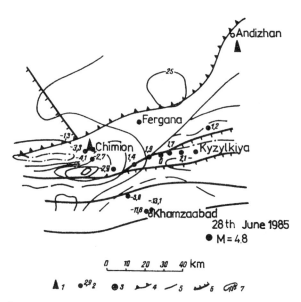

Figure 65. Localization of a source of precursing magnetic field anomaly at Chimion Station: (1) stationary points (MPP-1M); (2) auxiliary points of repeated magnetic measurements; (3) epicentre of Khamzaabad earthquake of 28th June 1985 with $M = 4.8$; (4) south Fergana flexural-fault zone; (5) deep-reaching fault; (6) activated fault segments; (7) isodynamic line.

epicentre of the expected earthquake Muminov established an auxiliary network in submeridional and latitudinal directions (Fig.65) on 19th-20th May 1985. One month later (20th-21st June) the measurements were repeated at the additional points, and the highest field changes (up to 13–18 nT) were identified south of Chimion.

Basing on the spatial distribution of the anomalous field changes at stationary stations and auxiliary points, with inclusion of the duration and form of the 25th June 1985 anomaly, Muminov predicted the place, intensity and time of the expected earthquake: 'south of Chimion Station at a distance of up to 100 km in 10–15 days an earthquake with $M = 5.0$ is expected'. Indeed, the Khamzaabad earthquake with $M = 4.8$ occurred on 28th June 1985 at the distance of 80 km SE of Chimion Station. As can be seen, the prediction of the three parameters was fully satisfactory.

Closure

Results of laboratory experiments and theoretical computations have been confirmed by prototype modelling of crustal processes under natural conditions. Under conditions of variable pressures the magnetic field varies due to not only elastic stresses but also other factors, including electrokinetic ones — which exemplifies the complexity of experimental studies and the necessity of accurate interpretation.

Changes of internal origin with characteristic times of 10–30 years appear in the Earth's magnetic field. By characteristic dimensions, they can be categorized as planetary and regional. The planetary changes are associated with processes in the Earth's core, and their intensity is counted in tens of nanoteslas.

Slow regional variations are associated with processes in the Earth's crust and have confined dimensions — from tens to hundreds of kilometres, and limited intensity — up to tens of nanoteslas. In investigations of mesoscale magnetic field changes of the Earth, field studies were conducted in three geodynamical test areas of Uzbekistan-Tashkent, Kyzyl-Kum and Fergana, by the methods of repeated route and area surveying and stationary round-the-clock magnetic observations. It has been possible to identify numerous magnetic effects having different characteristic times, nature and apparently dimensions of sources. They have been compared with earthquakes, regional geodynamical processes and processes at the earthquake focus.

Magnetic field changes with an intensity from units to first tens of nanoteslas and characteristic times from several months to years (mesoscale changes) are usually associated with incipient crustal earthquakes (the generation of earthquakes). Parameters of these factors are correlated with earthquake parameters. The longer the magnetic effect the stronger the intensity of the expected earthquake. A correlation also exists between the linear dimensions of the display zone of anomalous field change versus intensity of earthquake. We have provided quantitative relationships for these correlations.

Fast (short-term) changes in the magnetic field, with characteristic times from days to 2–3 weeks have the magnitude from units to 10–20 nT. The parameters of fast magnetic field changes remain uncorrelated with earthquake parameters.

The magnetic effects detected in Uzbekistan are of piezomagnetic, electrokinetic and electric-current origin, as elucidated in detail in respective chapters of this book.

The research on pulsed electromagnetic radiation of the Earth's crust has displayed numerous effects in pulsed electromagnetic radiation associated with earthquakes. By characteristic times of display they are correlated with short-term magnetic effects. The character of the anomalous effects in pulsed electromagnetic and magnetic fields seems to be unique, i.e. mechano-electrical. These effects are associated with the last stage of earthquake growth, i.e. generation of major fissures at the focus.

On the basis of the executed investigations a network of regular magnetic stations has been established in Uzbekistan test areas to provide an operational system of data acquisition for earthquake prediction. A technique of geomagnetic data processing in combination with other precursors has been worked out to secure the forecast of the place, time and intensity of earthquakes.

Using the above data and procedures the Forecast Service predicted 48 earthquakes in the years 1982–1985 (written evidence).

The prediction was verified in 70% cases for three parameters of earthquakes, 80% for two parameters and 90% for one parameter (time).

References

Abdullabekov, K.N. 1972. Investigation of geomagnetic field variations associated with elastic stresses in the Earth's crust of seismic regions (in Russian). Ph.D.Thesis Abstract. Moscow.

Abdullabekov, K.N. 1979. Seismicity of Kopetdag, Pamir, Tien-Shan and Caucasus versus solar activity (in Russian). *Uzb. geol. zhur.* 4: 6-9.

Abdullabekov, K.N., Ye.B.Berdilayev, A.N.Pushkov & V.A.Shapiro 1977. Preliminary results of experimental investigations in Charvak Reservoir (in Russian). In *Isledovaniye prostranstvienno-vremiennoy struktury geomagnitnogo pola*: 69-102. Nauka, Moscow.

Abdullabekov, K.N., Ye.B.Berdaliyev, A.N.Pushkov & V.A.Shapiro 1979. Local changes in the geomagnetic field upon impoundment of reservoir (in Russian). *Geomag. aeron.* XIX(2): 317-322.

Abdullabekov, K.N., L.S.Bezuglaya, V.P.Golovkov & Yu.P.Skovorodkin 1972. On the possibility of using magnetic methods to study tectonic processes. *Tectonophys.* 14(3/4): 257-262.

Abdullabekov, K.N., M.B.Gokhberg, G.A.Mavlanov et al. 1982. In *Electromagnetic radiation (electromagnetic forerunners of earthquakes)* (in Russian): 14-24. Nauka, Moscow.

Abdullabekov, K.N. & V.P.Golovkov 1970. Detection of seismomagnetic effect at Tashkent (in Russian). *Geomag. aeron.* 10(7): 1132-1133.

Abdullabekov, K.N. & V.P.Golovkov 1974. Geomagnetic field changes and crustal processes (in Russian). *Izv. AN SSSR. ser. fiz. zem.* 3: 93-100.

Abdullabekov, K.N., R.N.Ibragimov, G.A.Mavlanov et al. 1980. Geological and geophysical criteria of seismicity (in Russian). In *Seismicheskoye rayonirovaniye SSSR*: 169-174. Nauka, Moscow.

Abdullabekov, K.N. & S.Kh.Maksudov 1975. *Geomagnetic field variation of seismic regions* (in Russian). Fan, Tashkent.

Abdullabekov, K.N., S.Kh.Maksudov & M.Yu.Muminov 1981. Investigations on seismomagnetic effect in eastern Fergana (in Russian). In Second All-Union Symposium, *Stationary geomagnetic field magnetism and palemagnetism*, Part I: 92. Tbilisi.

Abdullabekov, K.N., M.Yu.Muminov & V.A.Shapiro 1979. Preliminary results of geomagnetic investigations on Fergana geodynamic test area (in Russian). *Uzb. geol. zhur.* 2: 26-29.

135

Abdullabekov, K.N. & F.D.Narmirzayev 1976. Preliminary results for secular geomagnetic field anomalies in Kyzyl-Kum area. In *Glavnoye geomagnitnoye pole i problemy paleomagnetizma*, Part I. Moscow.

Abdullabekov, K.N., V.A.Shapiro, Ye.B.Berdaliyev, M.Yu.Muminov, F.D.Narmirzayev & V.A.Narmatov 1984. In *Magnetometric investigations. 1976 Gazli earthquake. Geological and geophysical character of focuses* (in Russian): 168-183. Nauka, Moscow.

Abdullabekov, K.N. & R.I.Sultanbekov 1978. Role of electrokinetic phenomena in generation of some local geomagnetic field changes (in Russian). In *Rayonirovaniye seismicheskoy opastnosti i poiski predvestnikov zemletryaseniy*: 60-67. Fan, Tashkent.

Abdullabekov, K.N., A.A.Vorob'ev, M.B.Gokhberg et al. 1980. *Field investigations of disturbances in electromagnetic field of the Earth* (in Russian). Preprint No 8. IFZ AN SSSR, Moscow.

Akhmegzhanov, M.A., O.M.Borisov, R.N.Ibragimov & D.Kh.Yakybov 1971. Fault disturbances. In *Tashkent earthquake of 26th April 1966* (in Russian): 259-272. Fan, Tashkent.

Aki, K. 1956. Some problems in statistical seismology. *Zisin (J.Seismol. Soc. Japan)* 8: 205-208.

Aki, K. 1968. Statistcs of earthquakes in time. In *Earthquake prediction* (in Russian): 125-127. Mir, Moscow.

Akopyan, Ts.G. et al. 1974. Investigations on properties of local changes in geophysical fields of seismic regions in Armenia aimed at identification of earthquake precursors (in Russian). *DAN Arm. SSR* 59(2): 84-88.

Allen, C.R., M.G.Bonilla, W.F.Brace et al. 1975. Earthquake research in China. *Trans. Amer. Geophys. Union* 56: 838.

Altgauzen, N.M & O.M.Barsukov 1970. Temporal variations of electrical conductivity. In *Physical background for earthquake prediction methods* (in Russian): 104-110. Nauka, Moscow.

Ananin, I.V. 1977. *Seismicity of the northern Caucasus* (in Russian). Nauka, Moscow.

Anonymous 1956. *Electrokinetic properties of capillary systems.* Nauka, Moscow.

Anonymous 1968. *Earthquake forecast* (transl. from English). Mir, Moscow.

Anonymous 1971. *Tashkent earthquake.* Fan, Tashkent.

Anonymous 1976. *Earthquake forecast* (in Russian). Fan, Tashkent.

Anonymous 1982. *Electromagnetic forerunners of earthquakes.* Fan, Tashkent.

Anonymous 1982a. *Lithosphere of Pamir and Tien-Shan.* Fan, Tashkent.

Anonymous 1983. *Electric and magnetic forerunners of earthquakes.* Fan, Tashkent.

Aprodov, V.A. 1965. *Neotectonics, volcanic provinces and large seismic belts of the world* (in Russian). MGU, Moscow.

Argan, E. 1935. *Tectonics of Asia* (translation from French into Russian). ONTI, Moscow-Leningrad.

Aripova, F.M., N.T.Karimov, M.F.Inoyatov & A.A.Tadzhibayev 1975. Some results of measurements of stress condition in rock massifs (in Russian). *Uzb. geol. zhur.* 3: 59-61.

Arkhangelskiy, A.D. 1941. *Geological structure and geological history of the USSR* (in Russian). Gostoptekhizdat, Moscow-Leningrad.

Arkhangelskiy, A.D. & V.V.Gedynskiy 1936. Geological results of gravimetric investigations in Middle Asia and south-west Kazakhstan (in Russian). *Izv. AN SSSR, ser. geol.* 1.

Arora, B.R. & B.P.Singh 1979. Geomagnetic field precursors associated with earthquakes. *Mausam* 30(2-3): 317-322. 156.

Barsukov, O.M., L.S.Bezuglaya, V.N.Vadkovskiy & Yu.P.Skovorodkin 1968. On the nature of one of secular anomalies in the magnetic field of the Earth (in Russian). *Izv. AN SSSR, ser. fiz. zem.* 9.

Barsukov, O.M. & Yu.P.Skovorodkin 1966. Magnetic observations at Medeo fault (in Russian). *Izv. AN SSSR, ser. fiz. zem.* 5: 68-69.

Beahn, T.J. 1976. Geomagnetic field gradient measurements and noise reduction techniques in Colorado. *J.Geophys. Res.* 81(35): 6276-6280.

Belousov, V.V. 1962. *Fundamentals of geotectonics* (in Russian). Gosgeoltekhnizdat.

Belousov, V.V. 1984. Tectonic activity and transformation of the Earth's crust (in Russian). In *Activated zones of the Earth's crust*. Nauka, Moscow.

Belousov, V.V. & M.V.Gzovskiy 1954. Tectonic conditions and mechanism of earthquake generation (in Russian). *Trudy Geofiz. AN SSSR* 25 (152): 25-35.

Berdaliyev, E.B., A.N.Pushkov & Yu.P.Tsvetkov 1980. Results of measurements of local changes in geomagnetic field on a self-contained station (in Russian). *Geomag. aeron.* 19(2): 311-314.

Bezrodnyy, E.M., R.N.Ibragimov & V.I.Ulomov 1974. Dynamics of the Earth's crust and intensity of tectonic movements in the near-Tashkent seismic region (in Russian). In *Seismotectonics of Middle Asia and Far East*: 5-14. Sovet. Radio, Moscow.

Bonchkovskiy, B.F. 1954. Change in atmospheric electrical potential gradient as one of possible earthquake precursors (in Russian). In *Trudy geofiz. inst. AN SSSR* 25 (152): 192-206.

Bondar, G.N. & A.N.Pushkov 1975. Potential discrimination and interpretation of local secular anomalies. In *Analysis of spatial and temporal structure of geomagnetic field* (in Russian): 26-32. Nauka, Moscow.

Borisov, A.A. 1964. Morphology of Mohorovicic surface and its structural importance (in Russian). *Sovetskaya geologia* 4.

Borisov, A.A. 1967. *Deep structure of the Soviet territory by geophysical data* (in Russian). Nedra, Moscow.

Borisov, A.A. & V.V.Fedynskiy 1964. *Geophysical characteristics of geosynclinal regions in the Middle Asia. Activated zones of the Earth's crust, recent tectonic movements and seismicity* (in Russian). Nauka, Moscow.

Bott, M. 1976. *Internal structure of the Earth* (in Russian). Mir, Moscow.

Breiner, S. 1964. Piezomagnetic effect at a time of local earthquake. *Nature* 202: 790-791.

Breiner, S. 1966. A magnetometer array for investigation of the piezomagnetic effect in seismically active area. *Proc. 2nd US-Japan Conf. Res. Earthquake Prediction Problems*: 22-23. New York.

Breiner, S. 1968. *Grouping of magnetometers in investigations of the piezomagnetic effect in seismic zones (earthquake prediction)* (in Russian). Mir, Moscow.

Breiner, S. & R.L.Kovach 1968. Local magnetic events associated with displacements along San-Andreas Fault (California). *Tectonoph.* 6(1): 69-73.

Bune, V.I. 1970. Problem of the prediction of the place and time of strong earthquake in south Tien Shan zone. In *Physical background for search methods of earthquake prediction* (in Russian): 64-83. Nauka, Moscow.

Bunz, V.I., V.G.Gitis, Yu.K.Pushkin et al. 1980. *Discrimination of zones of the most probable earthquake focus by combined geological and geophysical symptoms* (in Russian): 42-45. Nauka , Moscow.

Burtman, V.S. 1964. *Talaso-Fergana fault (Tien-Shan)* (in Russian). Nauka, Moscow.

Butakov, V.F., A.A.Vorob"ev, N.N.Tankina, V.V.Ivanov & T.N.Oleynikova 1979. *Investigations on some properties of pulsed electromagnetic field at 'Krymskaya' seismic station of Kamchatka and comparison with their counterparts in other seismic regions* (in Russian). Tomsk.

Butovskaya, Ye.M., Kh.A.Akabayev & M.G.Flenova 1971. Structure of the Earth's crust in east Uzbekistan and adjacent regions by seismological data. In *Deep structure of the Earth's crust of Uzbekistan by geological and geophysical investigations*: 9-27 (in Russian). Fan, Tashkent.

Byuss, Ye.I. 1948, 1952 & 1955. *Seismic conditions of Transcaucasus*, 3 vols (in Russian). Izd. Gruz. SSR. Tbilisi.

Chernyavskiy, B.A. 1930. Atmospheric-electrical and electro-tellurian phenomena during earthquakes (in Russian). *Sots. nauka tekhn.* 12: 26-26.

Chernyavskiy, E.A. 1955. Atmospheric-electrical precursors of earthquakes. In *Meteorology and hydrology of Uzbekistan* (in Russian): 7-12. AN Uz SSR, Tashkent.

Chigarev, N.V. 1980. Seismogenesis and block structure of the Earth's crust (in Russian). *DAN SSSR* 255(2): 313-317.

Chigarev, N.V. 1982. Properties of spatial distributions of focuses and their migrations in Middle Asia. In *Electromagnetic precursors of earthquakes* (in Russian): 61-63. Nauka, Moscow.

Davis, D. 1975. Earthquake prediction in China. *Nature* 258: 286-287.

Davis, P.M. & F.D.Stacey 1972. Geomagnetic anomalies caused by man-made lake. *Nature* 240: 348.

Demenitskaya, R.M. 1967. *The Earth's crust and mantle* (in Russian). Nauka, Moscow.

Deryagin, B.V. & N.A.Krotova 1948. *Adhesion* (in Russian). AN SSSR, Moscow.

Deryagin, B.V., N.A.Krotova & V.P.Smilga 1973. *Adhesion of solid bodies* (in Russian). Nauka, Moscow.

Drushits, V.V. & V.N.Vereshchagin 1974. *Geochronological tables* (in Russian). Mingeo SSSR, Moscow.

Fedotov, S.A. 1965. Regularities in the distribution of strong earthquakes in Kamchatka, the Kuril Islands and North-East Japan (in Russian). *Tr. IFZ AN SSSR* 36(203): 66-92.

Fedotov, S.A. 1968. Seismic cycle, possible quantitative seismic regionalisation and long-term seismic forecast. In *Seismic regionalisation of the USSR*: 121-150.

Fedotov, S.A., G.A.Sobolev, S.A.Boldyrev et al. 1976. Long-term and test short-term prediction of Kamchatka earthquakes. In *Search for earthquake forerunners* (in Russian): 49-61. Tashkent.

Finkelshteyn, M.I. & N.Sh.Kambarov 1979. On geomagnetic variations as earthquakes precursors (in Russian). *Geom. aeron.* 19(6): 1126-1128.

Fischer, R.A. 1950. Tests of significance in harmonic analysis. In *Contributions to mathematical statistics*: 53-59. Wiley, New York.

Fotiadi, E.E., G.I.Karataev, V.G.Kolmogorov et al. 1970. *Baykal geodynamical test area* (in Russian). Inst. geol. geof. SO AN SSSR, Novosibirsk.

Fujita, N. 1977. Magnetic change accompanying the Haicheng earthquake of February 4, 1975. *J.Geod. Soc. Jap.* 23: 123.

Fuzaylov, I.A. 1977. *Structure of consolidated crust of west submergence of Tien Shan* (in Russian). Tashkent.

Gamburtsev, G.A. & P.S.Veytsman 1957. Properties of the Earth's crust structure in north Tien-Shan by geoseismic probing data and comparison with geological, seismological and gravimetric data (in Russian). *Bull. soveta po seismologii AN SSSR* 3.

Gayskiy, Yu.N. 1970. *Statistical investigation of seismic conditions* (in Russian). Nauka, Moscow.

Gutenberg, R. & S.Richter 1932. Seismicity of the Earth. *Handbuch der Geophysik* Bd. IY, Berlin.

Gutenberg, B. & C.F.Richter 1948. *Seismicity of the Earth* (in Russian). UL, Moscow.

Gutenberg, B. & C.F.Richter 1956. Magnitude and energy of earthquake. *Estr. Ann. Geofis.* 9: 1.

Ibragimov, R.N. 1970. *Seismotectonics of Fergana depression* (in Russian), , Fan, Tashkent.

Ibragimov, R.N. 1978. *Seismic zones of central Tien-Shan* (in Russian). Fan, Tashkent.

Ibragimov, R.N. & K.N.Abdullabekov 1974a. Periodicity of strong earthquakes in the Caucasus (in Russian). *Uzb. geol. zhur.* 6: 16-19.

Ibragimov, R.N. & K.N.Abdullabekov 1974b. Periodicity of strong earthquakes in west Tien-Shan (in Russian). *Uzb. geol. zhur.* 4: 42-45.

Ibragimov, R.N., K.N.Abdullabekov & I.M.Makhakamgzhanov 1975. On the nature of fissuring in central Kyzylkums by geophysical methods In First All-Union Symp., *Engineering-Geological Foundations of Seismic Microregionalization* (in Russian): 68-69. Fan, Tashkent.

Ibragimov, R.N., D.Kh.Yakubov & M.A.Akhmedzhanov 1973. *Newest structures of central Kyzylkums and their seismotectonic properties* (in Russian). Fan, Tashkent.

Idarmachev, Sh.G. & O.M.Barsukov 1978. 'Dam-type' earthquakes and variations of electrical resistivity of rock massifs at Chirkey Reservoir (in Russian). *DAN SSSR* 2: 302-305.

Inoyatov, M.F. 1982. Investigations on stress condition in rock massif by ultrasonic method (in Russian). *Uzb. geol. zhur.* 1: 16-20.

Ivanov, N.A., V.A.Shapiro & Z.A.Borisov 1978. Secular anomalies in the Ural geodynamical area. In *Contemporary movements of the Earth's crust* (in Russian): 59-64. Novosybirsk.

Johnston, M.J.S. 1974. Preliminary results from a search for regional tectomagnetic effects in California and Western Nevada. *Tectonoph.* 23(3): 267-275.

Johnston, M.J.S., N.W.Myren, G.O'Hara & J.H.Rodgers 1975. A possible seismomagnetic observation on the Garlock faret. *Califor. Bull. Seismol. Soc. Amer.* 65(5): 1129-1132.

Johnston, M.J.S., B.E.Smith & R.O.Burford 1980. Local magnetic field measurements and fault creep observations on the San Andreas fault. *Tectonoph.* 64(1-2): 45-57.

Johnston, M.J.S. & F.D.Stacey 1969a. Transient magnetic anomalies accompanying volcanic eruptions in New Zealand. *Nature* 224(5226): 1289-1290.

Johnston, M.J.S. & F.D.Stacey 1969b. Volcano-magnetic effect observed on Mt Ruapehu, New Zealand. *J. Geophys. Res.* 74(27): 6541-6544.

Kalashnikov, A.G. 1954. Potentials of magnetometric methods in the problem of earthquake prediction (in Russian). *Tr. geofiz. int.* 25(152): 162-180.

Kapitsa, S.P. 1955. Magnetic properties of igneous rock subject to mechanical stresses (in Russian). *Izv. AN SSSR, ser. geofiz.* 6: 489-504.

Karmaykl, R.S 1976. *Seismomagnetism and seismic prediction (search for earthquake forerunners)* (in Russian): 93-96. Fan, Tashkent.

Karnik, V. 1969a. Comparison of seismic activity of European seismic zones (in Russian). *Izv. AN SSSR, ser. fiz. zem.* 7: 70-77.

Karnik, V. 1969b. Seismicity of the European continent. In *Earth's crust and upper mantle*. Washington.

Kato, J. 1939. Investigations of the changes in the Earth's magnetic field accompanying earthquakes or volcanic eruptions. *Sci. Rept. Tohoku. Imp. Univ.* 1(27): 1-100.

Kawasumi, H. 1970. Proofs of 69-year periodicity and imminence of destructive earthquake in southern Kwam-to district and problems in the countermeasures thereof. *Chigaku Zassi* 79: 115.

Khadzhiyev, T.Kh., Yu.A.Trapeznikov, K.N.Abdullabekov & Yu.P.Skovorodkin 1984. Magnetic effects of remote strong earthquake in Andizhan area (in Russian). *Uzb. geol. zhur.* 4: 26-29.

Khain, V.Ye. 1965. Regenerated (epiplatformal) orogenic belts and their tectonic character (in Russian). *Sovet. geol.* 7: 3-17.

Khain, V.Ye. 1973. *General geotectonics* (in Russian). Nedra, Moscow.

Kharambayev, I.Kh. (ed) 1977. *Earth's crust and upper mantle of Middle Asia* (in Russian). Nauka, Moscow.

Khusamiddinov, S.S. 1983a. Investigations on interference of electromagnetic field of the Earth due to earthquakes. In *Electrical and magnetic precursors of earthquakes* (in Russian): 56-62. Tashkent.

Khusamiddinov, S.S. 1983b. Methods for investigation of variations of natural pulsed electromagnetic field. In *Electrical and magnetic precursors of earthquakes* (in Russian): 62-72. Tashkent.

Khusamiddinov, S.S. & K.N.Abdullabekov 1983a. Relation of the strength of natural pulsed electromagnetic field to earthquakes. In *Electrical and magnetic precursors of earthquakes* (in Russian): 80-89. Tashkent.

Khusamiddinov, S.S. & K.N.Abdullabekov 1983b. Results of studies on temporal variations of parameters of natural pulsed electromagnetic field. In *Electrical and magnetic precursors of earthquakes* (in Russian): 72-82. Tashkent.

Kim, L.E. & A.A.Aralbayev 1978. Periodicity of strong earthquakes in Kirgizia. In *Regionalization of seismic hazards and search for earthquake forerunners* (in Russian): 17-22. Fan, Tashkent.

Kirillova, I.V. 1957. Periodicity of destructive earthquakes in the Caucasus and Turkey (in Russian). *DAN SSSR* 115(4): 771-773.

Kirillova, I.V. & A.A.Sorskiy 1960. Methodology of seismic charting in 1:1000000 by example of the Caucasus (in Russian). *Bul. Soveta seismol.* 8: 121-124.

Kondo, G. 1968. The variation of the atmosphere electric field at the time of the earthquake. *Memoires of the Kakioka magnetic observatory* 13(11).

Kormiltsev, V.V. & V.A.Shapiro 1979. Magnetic field of currents. In *Electrical and magnetic measurements of forced polarization* (in Russian): 38-44. UNTs AN SSSR, Sverdlovsk.

Kosminskaya, I.P. 1968. *Methods of deep seismic probing in the Earth's crust and upper mantle* (in Russian). Nauka, Moscow.

Kosminskaya, I.P., G.G.Mikhota & Yu.V.Tulina 1958. Structure of the Earth's crust in Pamiro-Alay zone by geoseismic probing data (in Russian). *Izv. AN SSSR. ser. geofiz* 10.

Kosygin, Yu.A. 1968. *Tectonics* (in Russian). Nedra, Moscow.

Kozlov, A.N., A.N.Pushkov, R.Sh.Rakhmatullayev & Yu.P.Skovorodkin 1974. Magnetic effects upon explosions in rock (in Russian). *Izv. AN SSSR, ser. fiz. zem.* 3: 66-71.

Kramarenko, G.U., M.Kurbanov & A.A.Yuvshanov 1969. Some results of studies on secular anomalies in the geomagnetic field in near-Tashkent fault (in Russian). *Izv. AN Turkm SSR, ser. fiz. tekh. khim. geol.* 53: 30-35.

Krestnikov, V.N. 1961. History of geological development of Pamir and adjacent regions of Asia in Mezozoic-Cenozoic (in Russian). *Sovet. geol.* 4 & 7.

Krestnikov, V.N. & I.L.Nersesov 1969. Tectonic structure of the Pamir and the Tien-Shan and its relation to Mohorovicic relief (in Russian). *Sovet. geol.* 11: 36-39.

Kriger, L.R. 1978. Seismicity of Issyk-Kul depression in long-term forecast terms. In *Regionalization of seismic hazards and search for earthquake forerunners* (in Russian): 10-16. Fan, Tashkent.

Kuznestova, V.G. 1969. Comparison of geomagnetic variations recorded at a number of Transylvania stations (in Russian). *Geomag. aeron.* 9(6): 1120-1123.

Kuznetsova, V.G. & A.I.Bilinskiy 1972. Possible identification of local secular anomalies (in Russian). *Geomag. aeron.* 12(5): 954-957.

Kuznetsova, V.G., A.I.Bilinskiy & I.M.Rudenskaya 1975. Selection of anomalous secular field variation (in Russian). *Geomag. aeron.* 15(4): 757-760.

Kuznetsova, V.G. & V.Ye.Maksimchuk 1979. Possible identification of electrical conductivity anomalies by repeated high-sensitivity observations of geomagnetic field (in Russian). *Geofiz. sb. AN USSR* 89: 31-35. Naukova dumka, Kiev.

Kuznetsova, V.G. & M.I.Melnichuk 1973. Magnetometric investigations in Carpathian geodynamical area. In *Proceedings of 9th conference on permanent geomagnetic field and paleomagnetizm* Part I:84-85 (in Russian).

Kulagin, V.K. & Kh.Z.Sirozheva 1977. Migration of seismic activity in zones of Gissaro-Kokshal and Ilyak fault (in Russian). *DAN Tadzh SSR* XX(8): 32-36.

Kuchay, V.K. et al. 1978. The contemporary tectonic movements in the zone of the Vakhsh overfault (in Russian). *DAN SSSR* 240(3): 673-676.

Kuchay, V.K. 1981. *Zonal orogensis seismicity* (in Russian). Nauka, Moscow.

Kurskeyev, A.K. 1977. *Geophysical characteristics of the Earth's crust of Kazahstan* (in Russian). Nauka, Alma-Ata.

Lamakin, V.V. 1966. Periodicity of Baykal earthquakes (in Russian). *DAN SSSR* 176: 410-413.

Lapina, M.I. 1953. Geomagnetism and seismic phenomena (in Russian). *Izv. AN SSSR, ser. geof.* 5: 393.

Laryonov, U.A. 1976. Temporal changes in anomalous magnetic fields in Baykal geody-namical area. In *Proceed. contemporary movements of the Earth's crust* (in Russian): 48-54. Novosibirsk.

Lomnitz, C. 1966. Statistical prediction of earthquakes. *Rev. Geophysics* 4: 377.

Lomnitz, C. 1974. *Global tectonics and earthquake risk*. Elservier, Amsterdam.

Lursmanashvili, O.V. 1973. Periodicity of strong earthquakes in the Caucasus (in Russian). *Izv. AN SSSR, ser. fiz. zem.* 2: 80-86.

Lyapunov, A.A. & S.M.Fandyushina 1950. Reccurrence of earthquakes (in Russian). *Izv. AN SSSR, ser. geogr. geof.* XIY(6): 547-552.

Magnitskiy, V.A. 1965. *Internal structure and physics of the Earth* (in Russian). Nedra, Moscow.

Makhkamdzhanov, I.M. 1979. Results of temporal measurements of β_k in Kyzylkum geodynamical area. In *Seismological investigations in Uzbekistan* (in Russian). Fan, Tashkent.

Maksimovskikh, S.I. & V.A.Shapiro 1976. Field proton magnetometer of high accuracy T-MP (in Russian). *Geomag. aeron.* 16(2): 389-391.

Maksudov, S.Kh., Yu.P.Skovorodkin & B.Akramov 1979. Possible local geomagnetic vari-ation for Tashkent earthquake of 1966. In *Proceedings of 9th conference on permanent geomagnetic field, magnetism and paleomagnetism*, Part II: 154-156 (in Russian). Baku.

Mavlyanov, G.A., G.F.Tetyukhin et al. 1966. Contemporary tectonic movements in the central Kyzylkums. In *Problems of contemporary movements of the Earth's crust* (in Russian). AN SSSR, Moscow.

Mavlanov, G.A., Yu.P.Tsvetkov, Ye.B.Berdaliyev & K.N.Abdulbbekov 1983. Geomag-netic field variations associated with seismic activity in Tashkent area (in Russian). *DAN SSSR* 270(1): 72-74.

Mavlanov, G.A. & V.I.Ulomov 1976. Identification of earthquake precursors in Uzbek-istan. In *Search for earthquake forerunners* (in Russian): 25-38. Fan, Tashkent.

Mavlanov, G.A., V.I.Ulomov, K.N.Abdullabekov et al 1979. Anomalous variations of geomagnetic field in east Fergana: forerunner of Alay earthquake on 2nd November 1978 (in Russian). *DAN SSSR* 246(2): 294-297.

Mavlanov, G.A., V.I.Ulomov, K.N.Abdullabekov & S.S.Khusamiddinov 1979. Variations of electromagnetic field parameters used for earthquake prediction (in Russian). *Uzb. geol. zhur.* 5: 11-15.

Mei Shi Yun 1960. Seismic activity of China (in Russian). *Izv. AN SSSR, ser. geof.* 3: 381-395.

Melkanovitskiy, I.M. 1965. Deep geological structure of the Soviet part of Tien-Shan by geophysical data. In *22nd session of International Geological Union, papers by Soviet geologist, Problem 2* (in Russian): 15-25. Nedra, Moscow.

Meshcheryakov, Yu.A. 1968. Cooperation of socialist scientists of eastern Europe in the field of contemporary movements of the Earth's crust. In *Contemporary movements of the Earth's crust* (in Russian): 11-39. VINITI, Moscow.

Mizutani, H. & T.Ishido 1976. A new interpretation of magnetic field variation associated with the Matsushiro earthquakes. *J.Geom.Geol.* 28(2).

Mizutani, H., T.Ishido, T.Yokokura & S.Onishi 1976. Electrokinetic phenomena associ-ated with earthquakes. *Geophys. Res. Letters* 3(7): 965.

Mogi, K. 1973. Relationship between shallow and deep seismicity in the western Pacific region. *Tectonoph.* 17: 1.

Mogi, K. 1976. Regularities in spatial and temporal distribution of strong earthquakes and earthquake forecast. In *Search for earthquake forerunners* (in Russian): 19-24. Fan, Tashkent.

Moore, G.W. 1969. Magnetic disturbances preceding the 1964 Alaska earthquake. *Nature* 5226: 224.

Müller, E.K. 1930. *Geol. Beitr. zur Geophysik* 44.

Morderfeld, B.Ye. & V.N.Berkhovskiy 1976. Geomagnetic precursors of strong earthquakes. In *Seismicity and deep structure of Siberia and Far East* (in Russian): 248-250.

Mushketov, D.I. 1933. Seismic regionalisation of Middle Asia (in Russian). *Tr. seismol. int.* 34(2): 1-26. Moscow.

Myachkin, U.I. 1978. *Processes of earthquake development* (in Russian). Nauka, Moscow.

Nagata, T. 1965. *Rock magnetism* (in Russian). Mir, Moscow.

Nagata, T.S. 1969. Tectonomagnetism. *Trans. Sci. Assembly of IUGG/IAGA. IAGA Bull.* 27: 12-43. Madrid.

Naomi, F. 1977. Magnetic change accompanying the Haicheng earthquake of February 4, 1975. *Sokuchu Gakkaisy (J.Geod. Soc. Jap.)* 2: 23.

Nazarov, A.G., N.Agamirzayev et al. 1976. Search for earthquake forerunners in the Caucasus. In *Search for earthquake forerunners* (in Russian): 164-171. Fan, Tashkent.

Nersesov, I.L., A.Nurmagambetov & A.Sydykov 1980. Long-term prediction of strong earthquakes in northern Tien Shan (in Russian). *DAN SSSR* 250(6): 1352-1355.

Nersesov, I.L., V.S.Ponomarev & Yu.K.Kuchay 1974. Peculiarities of the spatial distribution of seismic background. In *Search for earthquake forerunners in forecast areas* (in Russian): 119-131. Nauka, Moscow.

Nesmeyanov, S.A. & I.I.Barkhatov 1978. *Newest and seismogenetic structures of west Gissaro-Alay* (in Russian). Nauka, Moscow.

Nikolayev, N.I. 1952. New tectonic stage of the growth of the Earth's crust (in Russian). *Bul. MOIN otd. geol.* 27(3): 93-94.

Nikolayev, N.I 1962. *Neotectonics and its pronouncement in the structure and relief of the Soviet territory* (in Russian). Gosgeolttekhnichizdat, Moscow.

Nikolayev, P.N. 1977. Methods of statistical analysis of fissures and reconstruction of tectonic stress fields (in Russian). *Izv. vuzov ser. geol. razv.* 2: 103-115.

Nikolayev, P.N. 1978a. Properties of reccurrence graph for earthquakes and their geological dependence. In *Results of extensive geophysical investigation of zones of seismic hazards* (in Russian): 199-203. Nauka, Moscow.

Nikolayev, P.N. 1978b. Stress condition and deformation mechanisms in the Earth's crust of Alpine folding area (in Russian). *Izv. vuzov ser. geol. razv.* 11: 65-78.

Nikonov, A.A. 1975. Migration of strong earthquakes along largest fault zones of Middle Asia (in Russian). *DAN SSSR* 225(2): 306-309.

Nikonov, A.A. 1977. *Holocene and contemporary movements of the Earth's crust* (in Russian). Nauka, Moscow.

Nurmatov, V.A., K.N.Abdullabekov & V.P.Golovkov 1980. Random nature of seismicity in Caucasus (in Russian). *Uzb. geol. zhur.* 3: 74-77.

Nurmatov, K.A., K.N.Abdullabekov & V.P.Golovkov 1981. Spatial and temporal characteristics of seismicity in some regions of Alpine belt. In *Summ. 'Geological and geophysical methods of investigations in seismic regions'* (in Russian): 45-46. Soviet Academy of Sciencies, Frunze.

Nurmatov, U.A., K.N.Abdullabekov & V.P.Golovkov 1983. Correlation of crustal and sub-crustal earthquakes in Pamir-Khindukush and Balkan regions (in Russian). *Uzb. geol. zhur.* 1: 34-38.

Obruchev, V.A. 1948. Basic features of kinetics and plastics of neotectonics (in Russian). *Izv. AN SSSR, ser. geol.* 5: 13-24.

Oganesyan, S.R., A.N.Pushkov, A.Kh.Bugramyan & E.G.Geodakyan 1979. Temporal variation of geomagnetic field and seismicity at Azat Reservoir (in Russian). *Izv. AN Arm. SSR nauka zem.* 32(5): 72-79.

Orlov, V.P. 1958. Magnetic secular anomalies in Middle Asia (in Russian). *Izv. AN SSSR, ser. geof.* 11: 1245-1247.

Orlov, V.P. 1959. Magnetic secular anomalies in Middle Asia and their correlation with tectonics (in Russian). *Tr. NIZMIR* 15/25: 65-71.

Orlov, V.P. & V.P.Sokolov 1965. Secular geomagnetic field changes and their anomalies. In *Present and past magnetic field of the Earth* (in Russian): 66-76. Nauka, Moscow.

Ostashevskiy, M.G. & Yu.P.Skovorodkin 1979. Effect of bay-type geomagnetic variations on local magnetic field. *Geomag. aeron.* XIX(3): 538-542.

Pakhmatulin, Kh.A. & R.I.Sultanbekov 1976. On the role of electrokinetic phenomena in generation of magnetic anomaly above Poltoratsk gas storage (in Russian). *Uzb. geol. zhur.* 4: 13-16.

Panasenko, G.D. 1974. Newest movements and seismicity of the Baltic Shield. In *Seismotectonics of Alpine folding belt of southern USSR and some adjacent territories* (in Russian): 87-95. Nauka, Moscow.

Parkhomenko, E.N. 1965. Electrical properties of rock (in Russian). Nedra, Moscow.

Parkhomenko, E.I. & Yu.M.Martyshev 1975. Electric charging and luminescence of minerals upon deformation and destruction. In *Physics of earthquake focuses* (in Russian): 151-159. Nauka, Moscow.

Patalikha, E.I., A.K.Kurskeyev & G.Sh.Zhaksylkov 1977. Structural and geophysical regionalization of Chu-Iliy zone (in Russian). *Vestnik AN Kaz. SSR* 9: 51-55.

Petrushevskiy, B.A. 1955. *Ural - Siberian Epihercynian platform and Tien-Shan* (in Russian). AN SSSR, Moscow.

Petrushevskiy, B.A. 1960. Correlation between maximum earthquakes and geological situation (in Russian). *Bul. Sov. seismol.* 8: 28-35.

Petrushevskiy, B.A. 1964. Newest tectonic movements of continental Asia and seismo-geological situation in the regions of their appearance. In *Activated zones of the Earth's crust* (in Russian): 45-47. Moscow.

Ponomarev, V.S., Yu.M.Teyteltaum & N.V.Tretyakova 1976. Properties of spatial distribution of seismicity at places of possible intensive earthquakes. In *Investigations on physics of earthquakes* (in Russian): 169-184. Nauka, Moscow.

Pribytkov, N.V. 1973. Unusual thunderstorm before earthquake (in Russian) *Priroda* 4: 122.

Pudovkin, I.M. et al. 1970. On anomaly of secular course in Kamchatka I. In *Geomag. aeron.* (in Russian) 10(1): 170-173.

Pudovkin, I.M. et al. 1970. On anomaly of secular course in Kamchatka II. In *Geomag. aeron.* (in Russian) 10(1): 173-175.

Pudovkin, I.M. et al. 1973. On direct relationship between geomagnetic variations and earthquake (in Russian). *DAN SSSR* 208(5): 1074-1077.

Pudovkin, I.M. et al. 1975. Geomagnetic variations as possible means of earthquake prediction. In *Geological fields and seismicity* (in Russian): 100-106. Nauka, Moscow.

Pudovkin, I.M., V.S.Pavlov, B.P.Reshetov et al 1965. State of the art report on secular anomaly of geomagnetic factors in Kamchatka (in Russian). In *The present and the past of the magnetic field of the Earth*: 96-100. Nauka, Moscow.

Pudovkin, I.M. & A.A.Tanichev 1969. Investigations on anomalous secular course of geomagnetic field in volcanic areas (Example of Kamchatka). In *Volcanism and generation of mineral deposits in Alpine geosynclinal zone* (in Russian): 9-10. Lvov.

Pudovkin, I.M. & A.A.Tanichev 1970. On investigation of anomalous secular changes in the geomagnetic field (in Russian). In *Permanent geomagnetic field*, Part I: 84-88. IFZ AN SSSR, Moscow.

Pushkov, A.N, Yu.P.Rivin & V.N.Kwanenko 1973. Synchronous profile observations to identify secular anomalies (in Russian). *Geomag. aeron.* 13(3): 543-544.

Pustovitenko, B.G. & A.G.Kamenobrodskiy 1976. Regularities in migration of earthquake focuses in the Crimea. *Investigations in earthquake physics* (in Russian): 184-193. Nauka, Moscow.

Pyankov, V.A. & V.A.Shapiro 1977. Some aspects of Butkin anomaly in secular course of geomagnetic field (in Russian). In *Geomag. aeron.* 17(3): 548-550.

Raleigh, B., G.Bennett, H.Craig et al. 1977. Prediction of the Haicheng earthquake. *ECS* 58(5): 236.

Rantsman, E.Ya. 1970. Places of earthquake and morphostructure of rocky countries (in Russian). Nauka, Moscow.

Rezvoy, D.P. 1962. West Tien Shan transverse deep reaching fault (in Russian). *Vestnik Lvov. univ. ser. geol.* 1.

Rezvoy, D.P. 1964. On tectonic nature of the western part of Pamir-Himalaya-Kun-Lun neotectonic rise (in Russian). In *Tectonics of Pamir and Tien Shan*: 49-68. Nauka, Moscow.

Rikitake, T. 1966a. A differential proton magnetometer - a geomagnetic project under the 5-year plan for earthquake prediction research. *Bull. Res. Inst. Toyo Univ.* 44: 1167-1178.

Rikitake, T. 1966b. Elimination of non-local changes from total intensity values of the geomagnetic field. *Bull. Earthquake Res. Inst. Tokyo Univ.* 44(3): 1041-1070.

Rikitake, T. 1968. Geomagnetism and earthquake prediction. *Tectonoph.* 6(1).

Rikitake, T. 1977. Probability of a great earthquake to ocurr off the Pacific coast of Central Japan. *Tectonoph.* 42(1): 43-51.

Rikitake, T. 1979a. Classification of earthquake precursors. *Tectonoph.* 54(3-4).

Rikitake, T. 1979b. *Earthquake forecast* (in Russian). Mir, Moscow.

Rikitake, T. 1980. Earthquakes and its forecasting. *Geojournal* 2: 145-152.

Rikitake, T., Y.Yamazaki & Y.Hagivara 1966a. Geomagnetic and geoelectric studies of the Matsushiro earthquake swarm 1. *Bull. Earthquake Res. Inst. Tokyo Univ.* 44: 363-408.

Rikitake T., Y.Yamazaki, Y.Hagivara et al. 1966b. Geomagnetic and geoelectric studies of the Matsushiro earthquake swarm 2. *Bull. Earthquake Res. Inst. Tokyo Univ.* 44: 409-418.

Rikitake, T., Y.Yamazaki, M.Sawada, Y.Sasai, T.Yoshido, S.Uzawa & T.Shimomura 1967. Geomagnetic and geoelectric studies of the Matsushiro earthquake. Swarm 4. *Bull. Earthquake Res. Inst. Tokyo Univ.* 45: 89-107.

Rikitake, T., T.Yoshiro & Y.Sasai 1968. Geomagnetic noises and detectability of seismo-magnetic effect. *Bull. Earthquake Res. Inst. Univ. Tokyo* 46(1): 137-154.

Rikitake, T., T.Yukutake, Y.Yamazaki, M.Sawada, Y.Sasai, Y.Hagiwara, K.Kawada, T.Yoshino & T.Shimomura 1966. Geomagnetic and geoelectric studies of the Matsushiro earthquake. Swarm 3. *Bull. Earthquake Res. Inst. Tokyo Univ.* 44: 1335-1370.

Rokityanskiy, I.I. 1980. Electrical conductivity anomalies in the Earth's lithosphere. In *Geophysical, geological and calamitous natural phenomena. Geology of continental margins* (in Russian): 45-51. Nauka, Moscow.

Sabardini, R., M.Bonafede & E.Boschi 1978. A thermomagnetic elastic model of the earthquake source mechanism. *Nuovo cim.* C1(86): 522.534.

Sadovskiy, M.A. 1979. Natural lumpiness of rock (in Russian). *DAN SSSR* 247(4): 829-831.

Sadovskiy, M.A., S.Kh.Negmatullayev, I.L.Nersesov & Yu.P.Skovorodkin 1979. Tectono-magnetic investigations in Dyshanbin and Garm areas (in Russian). *DAN SSSR* 2(249): 326-328.

Sadovskiy, M.A., G.A.Sobolev & N.I.Migunov 1979. Changes in natural radiowaves due to strong earthquakes in the Carpathians (in Russian). *DAN SSSR* 225(2): 316-319.

Sadovskiy, M.A., G.A.Sobolev & N.I.Migunov 1980. Recording of electromagnetic radiation preceding the Carpathian earthquake on 4th March 1977 (in Russian). *Preprint IFZ AN SSSR* 8.

Samokhvalov, M.A. 1974. *Variation of natural pulsed electromagnetic field in Kamchatka, Shipunskiy spit* (in Russian). Politekh. Inst., Tomsk.

Savarenskiy, E.F. & F.I.Monakhov 1948. Use of asimuths and angles of outcrop of seismic radiation in interpretaton of observations (in Russian). *Tr. Geofiz. inst. AN SSSR* 1(128): 35-55.

Savarenskiy, E.F., F.S.Sadikov et al. 1972. Structure of the Earth's crust in some Middle Asian regions by dispersion of phase velocity of surface waves (in Russian). *Uzb. geol. zhur.* 5: 55-59.

Sekiya, H. 1971. On the seismic activity in the southern part of the Kanto. *Q.J. Seismol. Japan Meteorol. Agency* 36: 13.

Shapiro, V.A. 1966. Seismomagnetic effect (in Russian). *Izv. AN SSSR, ser. fiz. zem.* 8: 61.

Shapiro, V.A. 1976. Local secular anomalies in geomagnetic field and the problem of earthquake prediction. In *Search for earthquake precursors* (in Russian): 200-207. Tashkent.

Shapiro, V.A. & K.N.Abdullabekov 1978a. An attempt to observe a seismomagnetic effect during the Gazli 17 May 1976 earthquake. *J. Geomagn. Geoelect.* 30(5).

Shapiro, V.A. & K.N.Abdullabekov 1978b. Observations on variations of magnetic field during Gazli earthquake on 17th May 1976. In *Geom. aeron.* 18(1): 177-179.

Shapiro, V.A. & N.A.Ivanov 1973. High-accuracy geomagnetic surveying of T on central Ural geodynamical test area in search for secular anomalies. In *Contemporary movements of the Earth crust* (in Russian). Tartu.

Shapiro, V.A., A.N.Pushkov, K.N.Abdullabekov, E.B.Berdaliyev & M.Muminov 1978. Geomagnetic investigation in the seismoactive regions of Middle Asia. *J. Geomagn. Geoelec.* 30(5).

Shebalin, N.V. & N.V.Kondorskiy (eds) 1977. *New catalogue of strong earthquakes in the USSR* (in Russian). Nauka, Moscow.

Shevtsov, G.I., N.I.Migunov, G.A.Sobolev & E.V.Kozlov 1975. Electric charging of feldspar upon deformation and destruction (in Russian). *DAN SSSR* 225(2): 313-315.

Sholz, C.M., L.R.Sykes & Y.P.Aggarwal 1973. Earthquake prediction: a physical basis. *Science* 181(4102): 803.

Sholz, C.M. 1977. A physical interpretation of the Haicheng earthquake prediction. *Nature* 267(5607): 121-124.

Shultz, S.S. 1964. Geostructural areas and position in the Earth's structure of orogenic zones by data of newest Soviet tectonics. In *Activated zones of the Earth's crust* (in Russian): 31-34. Nauka, Moscow.

Shuster, A. 1897 On lunar and solar periodicities of earthquakes. *Proc. R. Soc. Lond. Ser. A* 61: 455.

Sidorin, A.Ya. 1980. Earthquake forerunners in the NW part of Pacific seismic belt. *Volcan. seism.* 4: 88-98.

Sirozheva, Kh.Z. & V.K.Kulagin 1978. On migration of earthquakes and energy release in zones of major faults of Dushanbin geophysical area. In *Regionalisation of seismic hazards and search for earthquake forerunners* (in Russian): 3-9. Tashkent.

Sitdikov, B.B 1976. *Recent tectonics of Central Kyzylkums* (in Russian). Tashkent.

Skovorodkin, Yu.P. 1969. Magnetic investigations in the epicentral zone. In *Magnetism of rock and paleomagnetism* (in Russian). IFZ AN SSSR, Moscow.

Skovorodkin, Yu.P. 1980. Tectomagnetism and local magnetic variations in seismic zones. Ph.d. Thesis (in Russian). Moscow.

Skovorodkin, Yu.P. & L.S.Bezuglaya 1979. *Relation of geomagnetic variations to hydraulic conditions (example of Garm area)* (in Russian). Preprint No 1.

Smith, B.E. & M.J.S.Johnston 1976. A tectonomagnetic effect observed before a magnitude 5.2 earthquake near Hollister, California. *J.Geophys. Res.* 82(20): 3556-3560.

Sobakar', G.T., V.I.Somov & V.G.Kuznetsova 1975. Contemporary dynamics and structure of the Earth's crust in the Carpathians and adjacent territories (in Russian). Naukova dumka, Kiev.

Sobolev, G.A. 1973. Prospective operational prediction of earthquakes by electro-tellurian observations. In *Earthquake forerunners* (in Russian). VINITI, Moscow.

Sobolev, G.A. 1975. Application of piezoelectrical phenomena in geophysics. Ph.D. Thesis (in Russian).

Sobolev, G.A., V.N.Bagaevskiy, R.A.Lemastuyeva, N.I.Migunova & A.A.Khroma 1975. Study on mechanoelectrical processes in seismic regions. In *Physics of earthquake focus* (in Russian): 184-223. Nauka, Moscow.

Sobolev, G.A. & V.M.Demin 1980. *Mechanoelectrical phenomena in the Earth* (in Russian). Nauka, Moscow.

Sobolev, G.A. & V.N.Morozov 1970. Local interference of electric field in Kamchatka and its relation to earthquakes. In *Physical background for search of earthquake forecast methods* (in Russian): 110-121. Nauka, Moscow.

Stacey, F.D. 1964. The seismomagnetic effect. *Pure and Appl. Geophys.* 58(2): 5-22.

Stacey, F.D. 1965. The volcano-magnetic effect. *Pure and Appl. Geophys.* 62(3): 96-104.

Stacey, F.D. & P.Westcott 1965. Seismomagnetic effect limit of observability imposed by local variations in geomagnetic disturbances. *Nature* 206: 1209-1211.

Suyehiro, S. & H.Sekiya 1972. Foreshocks and earthquake prediction. *Tectonoph.* 14: 219.

Tal'-Virskiy, B.B. 1964. Some features of tectonic development of epiplatformal orogenic area of west Tien Shan. In *Activated zones of the Earth's crust, newest tectonic movements and seismicity* (in Russian): 109-122. Nauka, Moscow.

Tal'-Virskiy, B.B. 1972. *Tectonics and geophysical fields of oil and gas deposits in the central Middle Asia.* Ph.D.Thesis (in Russian). Moscow.

Tal'-Virskiy, B.B. & I.A.Fuzaylov 1970. Chart of magnetic anomalies in central Middle Asia. Moscow.

Tamrazyan, G.P. 1962. Periodicity of seismic activity in recent 1500-2000 years (in Russian). *Izv. AN SSSR, ser. geofiz.* 1: 76-85.

Tamrazyan, G.P. 1978. Major spatial and temporal regularities in the seismotectonic development of the Earth in 20th century (in Russian). *Izv. AN Arm SSR nauk. zem.* 31(4): 17-31.

Tazima, M. 1966. Accuracy of recent magnetic survey and a locally anomalous behaviour of the geomagnetic secular variation in Japan. *Theses, Tokyo Univ.* 133.

Tazima, M. 1968. Accuracy of recent magnetic survey and locally anomalous behaviour of the geomagnetic secular variation in Japan. *Bull. Geogr. Survey Inst.* 13(2): 343-367.

Tikhonov, A.N., A.G.Ivanov, V.A.Troitskaya & B.P.Dyakonov 1974. On relation of tellurian currents and earthquakes (in Russian). *Tr. geofiz. inst. AN SSSR* 25(152): 181-191. Moscow.

Tserfas, K.E. 1971. Phenomena of atmospheric electricity preceding earthquakes. In *Tashkent earthquake of 26th April 1966* (in Russian): 184-186. Tashkent.

Tsubokawa, I. 1969. On relation between duration of crustal movement and magnitude of earthquake expected. *J.Geod. Soc. Japan* 15(75).

Tsubokawa, I. 1973. On relation between duration of precursory geophysical phenomena and duration of crustal mocement before earthquake. *J.Geod. Soc. Japan* 19(116).

Tsvetkov, Yu.P., V.V.Polyakov, B.P.Murashov & A.F.Fedosova 1977. PN-001 proton magnetometer. In *Geomagnetic apparatus* (in Russian): 3-8. Nauka, Moscow.

Turner, H.H. 1925. Note on the 284-year cycle in Chinese earthquake. *Monthly notices. Roy. Astron. Soc. Geophys. Suppl.* 1(6).

Ulomov, V.I. 1959. Regional section of the Earth's crust in Middle Asia and pump yield at Tashkent (in Russian). *Izv. AN Uz SSR, ser. fiz-mat.* 2.

Ulomov, V.I. 1960. Some properties of the structure of the Earth's crust in Middle Asia by records of strong explosion (in Russian). *Izv. AN SSSR, ser. geof.* 1: 131-134.

Ulomov, V.I. 1962. Results of investigations of the deep structure of the Earth's crust in Middle Asia by seismological data (in Russian). *Izv. AN SSSR, ser. geofiz.* 2: 1307-1319.

Ulomov, V.I. 1966. *Deep structure of the Earth's crust in south-east Middle Asia by seismological data* (in Russian). Tashkent.

Ulomov, V.I. 1971. Luminous electrical phenomena accompanying earthquake. In *Tashkent earthquake of 26th April 1966* (in Russian): 181-184. Tashkent.

Ulomov, V.I. 1972. *Attention, earthquake!* (in Russian). Tashkent.

Ulomov, V.I. 1974. *Dynamics of the Earth's crust and earthquake prediction* (in Russian). Tashkent.

Ulomov, V.I. 1982. Electric charges inside the Earth. In *Electromagnetic earthquake precursors* (in Russian). Nauka, Moscow.

Ulomov, V.I. & B.Z.Mavashev 1971. Forerunner of Tashkent earthquake. In *Tashkent earthquake of 26th April 1966* (in Russian): 188-191. Tashkent.

Undzenkov, B.A. & V.A.Shapiro 1967. Seismomagnetic effects in magnetite ore deposits (in Russian). *Izv. AN SSSR, ser. fiz. zem.* 1: 121-126.

Usami, T. & S.Hisamoto 1970. Future probability of a coming earthquake with intensity V or more in the Tokyo area. *Bull. Earthquake Res. Inst. Univ. Tokyo.* 48: 331.

Usami, T. & S.Hisamoto 1971. Future probability of a coming earthquake with intensity V or more in the Kyoto Area. *Bull. Earthquake Res. Inst. Univ. Tokyo* 49: 115.

Usmanova, M.T. 1979. Brittle destruction crack as one of possible sources of pulsed electromagnetic radiation of rock under natural conditions (in Russian). *Uzb. geol. zhur.* 2: 30-33.

Utsu, T. 1972. Aftershocks and earthquake statistics. *J. Fac. Sci. Hokkaido Univ.* 7(4): 1.

Vadkovskiy, V.N., V.A.Lyakhovskiy & Yu.S.Tyupkin 1978. *Temporal evolution of seismic activity of the Balkan region (algorithms and results of data processing)*: 11-23 (in Russian). Izd. AN SSSR.

Vasul, V. 1974. A summary of studies on luminous phenomena accompanied earthquake. *Dokkyo Medical Univ. Japan.*

Vilkovich, Ye.V., V.I.Keilis Borok, V.M.Podgayetskaya, A.G.Prozorov & Ye.Ya.Rantsman 1974. Statistical analysis of earthquake catalogues and morphostructure of east Middle Asia. In *Theoretical and computational geophysics* (in Russian) 1: 69-117. Moscow.

Vilkovich, Ye.V., A.G.Prozorov & D.A.Kadson 1976. Correlation of earthquakes. In *Search for earthquake precursors* (in Russian): 132-140. Tashkent.

Volarovich, M.P. & Ye.I.Parkhomenko 1954. Piezoelectric effects in rock (in Russian). *DAN SSSR* 99(2): 239.

Volarovich, M.P. & Ye.I.Parkhomenko 1955. Piezoelectric effects in rock (in Russian). *Izv. AN SSSR, ser. geofiz.* 2: 215-222.

Volvovskiy, I.S., V.Z.Ryaboy & V.I.Shraybman 1962. Deep geological structure of Fergana depression by geophysical data (in Russian). *Sovietskaya geologia* 1: 156-160.

Volvovskiy, B.S., I.S.Volvovskiy, B.B.Tal'-Virskiy et al 1973. Structure of the Earth's crust in east Middle Asia (in Russian). *Geofiz. Biul* 26: 52-64.

Vorob"ev, A.A 1970. Possible electric charging inside the Earth (in Russian). *Geol. geofiz.* 12: 3-6.

Vorob"ev, A.A. 1973. *Energy transformations in the Earth's crust* (in Russian). Izd. Politekh. Inst., Tomsk.

Vorob"ev, A.A 1977. *Mechanoelectrical processes and energy transformations upon plastic deformation* (in Russian). Politekh. Inst., Tomsk.

Vorob"ev, A.A. 1978. *Evaluation of efficiency of mechanoelectrical transformations in rock under natural conditions* (in Russian). VINITI (Dep.28 dek. 3832-79).

Vorob"ev, A.A. 1979a. *Electric apparatus and electric insulation* (in Russian). Energiya, Moscow.

Vorob"ev, A.A 1979b. *Forecast of earthquake instant by parameters of natural pulsed electromagnetic field* (in Russian). VINITI (Dep. No 1779-79).

Vorob"ev, A.A 1979c. *Natural geoelectrical and electromagnetic fields and their applications* (in Russian). VINITI (Dep. 12 febr. 648-79).

Vorob"ev, A.A., M.A.Samakhvalov, A.F.Gorelkin, R.N.Ibragimov, M.T.Usmanova & A.N.Khogzhaev 1976. Anomalous changes of intensive natural pulsed electromagnetic field at Tashkent before earthquake (in Russian). *Uzb. geol. zhur.* 2: 9-11.

Vvedenskaya, A.V. 1969. *Investigations of stresses and faults in earthquake focuses in terms of the theory of dislocation* (in Russian). Nauka, Moscow.

Whitcomb, I.H., I.D.Garmany & D.Anderson 1973. Earthquake prediction of seismic velocities before the San Fernando earthquake. *Science* 180: 632.

Yakubov, D.Kh., M.A.Akhmedzhanov & O.M.Borisov 1976. *Regional faults in the central and southern Tien Shan* (in Russian). Tashkent.

Yamazaki, V. 1977. Tectonoelectricity. *Geophys. surv.* 3(2): 123-142.

Yanogikhara, K. & T.Posimatsu 1968. Local changes in tellurian currents at Kakioke before earthquake. In *Earthquake prediction* (in Russian): 137-138. Mir, Moscow.

Yanshin, A.L. 1948. Methods of investigations on ridge folding structure by example of the Ural, Tien-Shan and Mangyshlak (in Russian). *Izv. AN SSSR, ser. geol.* 5: 135-154.

Yanshin, A.L. 1965. Tectonic structure of Euroasia. *Geotectonics* 5: 7-35.

Yasuo, Y. 1968. A study on the luminous phenomena accompanied with earthquakes. *Memoirs of the Kokioka magnetic observatory. Japan* 13(1): 25-61.

Yavorskiy, B.M. & A.A.Detlaf 1964. *Physical manual* (in Russian). Moscow.

Yermilin, V.I. & N.V.Chigarets 1981. *The orogenesis and seismicity of Pamir-Alay* (in Russian). Nauka, Moscow.

Yerzhanov, Zh.S., A.K.Kurskeyev, M.P.Rudina, Ye.Alipbekov, V.V.Kazakov, Z.M.Nascyrova & T.Ye.Nysanbayev 1976. The techniques of investigations on seismomagnetic effects in Alma-Ata area and preliminary results (in Russian). *Izv. AN Kaz. SSR, ser. geolog.* 6: 77-81. Nauka, Alma-Ata.

Yudakhin, F.N. 1983. *Geophysical fields, deep structure and seismicity of the Tien Shan* (in Russian). Ilim, Frunze.

Yur'yev, A.A. 1967. On neotectonics of the west border of the Turkestan-Zarafshan mountain range. In *Tectonic movements and recent structures of the Earth's crust* (in Russian): 342-349. Nedra, Moscow.

Zhidkov, M.P. & V.G.Kosobokov 1978. The recognition of possible generation of strong earthquakes (intersection of east Middle Asia lines). In *Problems of earthquake prediction in the Earth's structure* (in Russian): 48-71. Nauka, Moscow.

Zubkov, S.I. & N.I.Migunov 1975. Instant of electromagnetic precursors of earthquakes (in Russian). *Geomag. aeron.* 15(6): 1070-1074.